A Taste
for the
Beautiful

A Taste
for the
Beautiful

||||||||||||||||||||||||||||||||||||||||||||||||||||||||||||||||||||||||||||||||||||||||||||

The EVOLUTION *of* ATTRACTION

||||||||||||||||||||||||||||||||||||||||||||||||||||||||||||||||||||||||||||||||||||||||||||

MICHAEL J. RYAN

PRINCETON UNIVERSITY PRESS

PRINCETON AND OXFORD

press.princeton.edu

Jacket design by Andrea Guinn
Jacket images courtesy of Biodiversity Heritage Library, Flickr

ISBN 978-0-691-16726-8

British Library Cataloging-in-Publication Data is available

This book has been composed in Adobe Caslon Pro
and F CaslonTwelve ITC for display

Printed on acid-free paper. ∞

Printed in the United States of America

1 3 5 7 9 10 8 6 4 2

In memory of *Stan Rand, fellow traveler*

In appreciation of the
Smithsonian Tropical Research Institute

Contents

||||||||||||||||||||||||||||||||||||||||||||||||||||||||||||||||||||||||||||||||||||||||||||||||||||||||||

Preface

||||||||||||||||||||||||||||||||||||||||||||||||||||||||||||||||||||||||||||||||||||||||||||

I AM A SCIENTIST. This occupation affords me the opportunity to try to understand a small part of the natural world. I am also a professor. This occupation allows me to explain to others my findings in the context of the fields in which I work, animal behavior and evolution. My audience is usually college students and other scientists, although I often present lectures to a more general audience and quite frequently explain what I do to friends and family who do not have a background in science. I usually find that regardless of their background, many folks are interested in these stories of how sexual beauty plays out in nature and how our experiments in the laboratory give us insights into how and why sexual beauty evolved. My purpose in writing this book is to share these stories of sexual beauty, some from my research but mostly from others, with a much wider audience.

I owe a huge debt of gratitude to a large number of people. Marc Hauser first suggested that I write this book, and he has been a tough and constructive critic throughout all stages of its preparation. A stay as a Fellow at the Institute for Advanced Study (WIKO) in Berlin afforded me the opportunity to plan this book, and during the planning stage I benefited greatly from the feedback of Karin Akre; Robert Trivers; Idelle Cooper; Doug Emlen; Eric Lupfer, my agent at William Morris Enterprises; and Alison Kalett, my editor at Princeton University Press. Marc Hauser and Idelle Cooper read the entire manuscript, and Alison Kalett and Karin Akre did the same and also provided detailed editorial comments.

During the later stages of writing this book, I had a bad accident that left me in a wheelchair with a spinal cord injury. I would not have been able to complete this work without the expert and generous care at St. David's Rehabilitation Hospital and Rehab Without Walls in Austin, Texas. I am eternally grateful to the staff of both of these organization, who I feel are part of a second family. During my time in the hospital, I received strong support and assistance on various tasks related to this book from Emma Ryan, Lucy Ryan, Marsha Berkman, Sofia Rodriguez, Idelle Cooper, Mirjam Amcoff, Fernando Mateos-Gonzalos, Karin Akre, Rachel Page, Caitlin Friesen, Tracy Burkhart, David Cannatella, May Dixon, Claire Hemingway, Ryan Taylor, and Kim Hunter. All of them are very special people.

My own research that I discuss in this book was funded by the National Science Foundation, the Smithsonian Institution, and the University of Texas. To each of these institutes I am most grateful.

A Taste
for the
Beautiful

ONE

||||||||||||||||||||||||||||||||||||||||||||||||||||||||||||||||||||||||||||||||||||||||||||||||||||||||||

Why All the Fuss about Sex?

The sight of a feather in a peacock's tail, whenever I
gaze at it, makes me sick!   —*Charles Darwin*

NATURE USUALLY GETS DOWN TO BUSINESS. Let's think about sleep. When I go to bed, I pull back the sheets, put my head on the pillow, and I'm in dreamland. I do not have a sleeping ritual, I don't dance, sing, chant, or perfume myself. I just go to sleep. So do most animals. Eating is the same way. When a howler monkey finds an edible leaf, she plucks it and eats it; a heron just throws back his head and swallows the fish he speared out of the water; and a cheetah doesn't do a celebratory dance before she starts to devour the gazelle she just brought down, even though she sprinted at her personal best of 75 mph to do so. Granted, in our own species we can sometimes make a bigger deal out of eating, especially when a meal coincides with a special event. But for the most part, we are little different from the howler, the heron, and the cheetah. Take a bite, give it a good chew, and gulp it down. Much of life for most animals is like that— the job is to just get it done.

Sex is different: a just-get-it-done policy won't get it done. In humans and most other animals, extensive courtship rituals precede the sex act. Most of our sexual rituals are laden with accessories, including candles and music, poems and flowers, and even special wardrobes. The list goes on, but it is no less diverse for animals. Animals sing and dance, they perfume themselves, they show off their colors and even light themselves up, all in the hope of attracting a mate. Although we distinguish ourselves in the language and technology we deploy in courtship, all animals have evolved spectacular, even obscene, morphologies and behaviors as both sexual lures and strategies for consummation. The colors of butterflies and fishes, the songs of insects and birds, the sexual odors of moths and mammals all evolved in the service of sex. The same is true for many of the traits in our own species that make women sigh and men gasp when someone of striking beauty crosses their paths. These aspects of sexual beauty evolved not because they make their bearers live longer but because they enable them to mate more and thus pass on more offspring and genes to the next generation.

Sexual beauty is everywhere, woven through the fabric of all sexually reproducing animals. We humans strive for beauty; we pay for it; we judge whether others have it; and if they do, we treat them better. Animals and humans both go to extreme lengths to appear beautiful to those who judge them. Peacocks evolve magnificent tails that cause peahens to sway, fishes sport bright colors that catch the eye of the other sex, crickets chirp endearingly to their mates, and spiders dance and vibrate their webs to show off. We humans take a more active role in engineering our beauty than do most other animals. Perfumes, fashion, cars, and music have all been employed in the service of sexual beauty, as have the surgeon's knife and a pharmacopeia of drugs. But to enhance one's beauty, either through the painstakingly slow process of evolution or the more immediate gratification of beauty-engineering, one must have some notion of what is beautiful.

This book is about sexual beauty, where it comes from and what it is for. Of course, many have been inspired to write in appreciation of natural beauty and the enchanting mating behaviors that occur in wild animals. Their emphasis is usually on the details of beautiful male traits: How does having such a long tail benefit the peacock? How many carotenoids does the male guppy need to eat to be so brilliantly orange?

How many syllables can a songbird pack into his complex vocal repertoire to make him even more sexy to females? These are interesting questions, but they represent only one-half of the equation of sexual beauty, because they ignore what is going on inside the head of those who actually judge beauty. Such studies often assume that the female brain must evolve tools to figure out what is beautiful. But instead, the converse is often true. The brain has a long evolutionary history that biases how it assesses the entire world around it, not just the world of sex; and it functions within the framework of numerous neurobiological and computational constraints. I argue that instead of the brain having to evolve to detect beauty, the brain determines what is beautiful, and all of its constraints and contingencies give rise to a breathtaking diversity of sexual aesthetics throughout the animal kingdom. In this book, I will show that to understand what beauty is, we need to understand the brain that perceives it.

I will expand our understanding of sexual beauty by asking how the details of an animal's brain give rise to its sexual aesthetics, which, in turn, drive the evolution of beauty in that species. Specifically, I argue that beauty only exists because it pleases the eyes, ears, or noses of the beholder; more generally, that *beauty is in the brain of the beholder*. Some of the brain's neural circuitry has evolved to sense and respond to sexual beauty so that animals can find a good mate. But the brain also has other things on its mind besides sex. Other adaptations of the brain, such as those that help an animal find food, avoid becoming food, or recognize the difference between its mother and its father, can have unintended but important consequences on how that brain defines beauty. Only when we understand the biological basis of sexual aesthetics can we understand how sexual aesthetics drive the evolution of sexual beauty.

I have a unique perspective to offer on these issues as I have spent the past forty years studying the sexual behavior of a tiny, bumpy frog in Central America.[1] This work has opened my eyes and mind to both the diversity of sexual behavior in the animal kingdom and a core unifying theory that I have developed called *sensory exploitation*. The key idea is simple: features of the female's brain that find certain notes of the males' mating call attractive existed long before those attractive notes evolved. Thus, females are the biological puppeteers, making the males sing exactly what their brains desire. Beauty is indeed in the brain of the beholder, and in most cases, that means the female's brain, although I will review numerous cases where males judge female beauty and where there is mutual display and assessment of beauty by both sexes. This simple idea contributed to a paradigm shift in the study of sexual selection, one in which the importance of the sexual brain as a driver of evolution finally was acknowledged.

In this chapter, I will give some background on how scientists have come to understand the evolution of beauty and also explain which sex usually evolves this beauty and why. In the next one, I'll focus on the bumpy frog that has been the focus of much of my scientific brain power, to show how scientists actually go about learning how the brain relates to mating behavior. Chapter 3 delves into how the brain defines beauty by discussing the evolution of sensory systems and the cognitive processing of sensory information. Chapters 4 through 6 describe what is known about visual, acoustic, and olfactory beauty throughout the animal kingdom. Chapter 7 describes some biological underpinnings to the claim that percepts of beauty are sometimes fickle. And in chapter 8, I describe how some percepts of beauty lie masked and unknown until just the right individual appears to elicit attraction. This logic is extended to provide an evolutionary understanding of how various human enterprises, from the fashion industry to pornography, have been able to exploit these hidden preferences. In the epilogue, I close the book with some comments about the biological basis of beauty.

In our search for answers about beauty, we will explore nature and journey to where scientists have studied some of the world's most stunningly beautiful animals. We will probe the basic premises of why sexual beauty had to evolve and delve into new findings from neuroscience that provide insights into how the brain perceives beauty. The analogies between animals and humans might cause us to rethink our own sexual aesthetics. As with much of biology, the best place to start thinking about sexual beauty is with Charles Darwin. Where I will depart from Darwin is within an arena that he knew little about: the brain.

* * *

It is hard to overestimate the impact of Charles Darwin's theory of evolution by natural selection on our view of humanity's place in the universe. It is one of the crowning intellectual achievements of humankind, ranking right up there with Copernicus's theory of celestial motion, Newton's laws of physics, and Einstein's theory of relativity. His book, *On the Origin of Species*, sold out in a few days; subsequent editions continued to sell out for decades; and it is still one of the most widely cited books in the world.[2]

The most amazing thing about natural selection is its brilliant simplicity, which can be unpacked into three ideas or principles. The first, which comes from Thomas Malthus's *Essay on the Principle of Population*, is that the rate of reproduction outstrips the available resources to support it—not all offspring survive to reproduce.[3] Consider a pair of house flies that sneak into your dwelling through a small tear in the window screen. This couple is capable of producing five hundred offspring during their short lifetime of one month. If all of their offspring and their future progeny survived to reproduce, six months later you would be inundated by about two trillion flies with a combined weight of more than 2,500 tons, whose body mass would cover more than one thousand square miles, an area close to the size of Luxembourg. Luckily, this doesn't happen, as most of these flies die, and only a handful survive.

The second principle is that differential survival is not always random. Some survivors are just lucky—for example, those who happen not to be around as your fly swatter comes bearing down. But others survive because they are "better"; they have adaptations that allow them to avoid your swat and live to reproduce. Perhaps they are more sensitive to the wind displacements caused by the fly swatter, or they have faster flight muscles that allow escape before they get splat. But they are survivors, and they get to stay on the island, or at least in your house.

The third principle is that if variation in survival traits has a genetic component, these traits will be differentially passed down to the next generation. If the surviving flies have genes for faster flight muscles, for example, so will their offspring. These offspring will constitute a new generation of flies that fly faster, live longer, and reproduce more. This is how natural selection causes the evolution of survival traits. Time to fix that tear in your window screen.

When Darwin, along with Alfred Wallace, formulated the theory of natural selection, he never suggested it explained everything—he never thought that every aspect of every individual was an adaptation for survival.[4] He was aware of the power of culture, in animals as well as humans. Darwin also understood random variation, which occurs when alternative forms of the same trait can become fixed in small populations. But one thing he did not understand, at least not immediately, was the peacock's tail. It caused him such consternation, he wrote to the botanist Asa Gray, that it made him sick. We know that Darwin was often sick, and a hypochondriac to boot, but such malaise in response to something so magnificent seems a bit extreme.[5] The peacock's tail is the mascot for scientific studies of animal beauty, but for Darwin it was a stark reminder of what his theory did not explain, and it motivated him to find a new theory to complement that of natural selection. He called it *sexual selection*.[6]

* * *

The peacock is a majestic and beautiful beast. He initiates courtship with a female by erecting his feathers to form a fan that spreads out more than 180 degrees. He has two hundred feathers up to four feet long that are adorned with eyelike spots and have an iridescent sheen that causes them to sparkle brilliantly in the sunlight. Once they are erect, he shakes, rattles, and rolls his feathers, causing them to hum like an engine and the eyespots to vibrate hypnotically. All of this beauty evolved in the service of sex. Peahens get to choose their mates, and peacocks evolved their beauty to better compete in the sexual marketplace, where only the beautiful get chosen to pass their genes forward.

A peacock displaying in all his splendor is a majestic sight to us and to peahens alike. But have you ever seen a peacock run or fly? It's pathetic! Dragging his tail behind him, he can't outrun a child let alone a fox, and he can barely fly. If Darwin was correct that natural selection causes adaptations for survival by weeding out the weak, where did this monstrosity come from, and why wasn't it culled out long ago? This is why a mere feather was so distressing to one of science's greatest minds. But it was mental, not physical, duress that caused this particular malady. The peacock's tail offered a major challenge to Darwin's theory of natural selection, so he went to work on another theory to explain how it could evolve.

The peacock's tail was not the only challenge to Darwin's calculus of survival evolution; it was just the tip of the iceberg. In his second-most famous book, *The Descent of Man and Selection in Relation to Sex*, published twelve years after *On the Origin of Species*, Darwin noted that many animals, not just peacocks, harbored traits that seemed at odds with the process of natural selection. Many of these traits also appear beautiful to us and seem superfluous to the animal's survival. Fireflies light up when they glide across a nocturnal meadow; crickets spend hours chirping during the summer nights; coral reef fishes sport colors that focus our gaze; frog choruses announce the coming of spring; canaries sing arias that have charmed their mates for millennia and humans for centuries; bowerbirds decorate and paint their bowers with such creativity that one researcher invoked a comparison to Matisse;[7] and Irish elk carried around eighty-eight-pound antlers with such high calcium demands that this might have eventually led to their extinction.[8] We are no more restrained with our sexual beauty, as we invest billions of dollars each year to paint, perfume, and trim parts of our bodies that make us more sexually attractive. None of this has anything to do with improved survival.

These nonsurvival traits share other commonalities. Most of them are more developed in males than females; they are usually employed in courtship or in battle for mates; and, as first haunted Darwin, many of these traits are detrimental to survival. Darwin called these *secondary sexual characters* because they differed between the sexes and were associated with reproduction, although not crucial for it. How they evolved required some additional theorizing.

Artificial selection provides some instructive examples of how these showy sexual characters might evolve. It might be one of the most important human inventions since the control of fire, and Darwin used artificial selection as an analogy to natural selection. In artificial selection, humans are the agents of selection. We decide which traits, as the targets of selection, will evolve to meet our predetermined goals. We often selectively breed organisms for utilitarian purposes, such as disease resistance in crops and greater meat yield in cattle. But we also breed animals to please our aesthetic senses. Fish hobbyists breed aquarium fishes with spectacular colors and even implant foreign genes to make some fish glow in the dark, and we are all familiar with breeds of domestic dogs that humans have engineered because they are cute rather than functional.

Based on his intuitions derived from artificial selection, Darwin reasoned that if female animals also had their own aesthetics, their own standards of beauty, they too could exert selection to enhance their species' beauty. If female canaries were attracted to more variable male song, males with more variable song would produce more offspring, and canary song would evolve to be highly variable over time. If female peacocks found longer feathers to be sexually beautiful, they would choose to mate with males that have longer feathers, and consequently those males would have more offspring. Longer tails would come to flourish in future generations, even if these tails increased the male's predation risk. A short-feathered peacock that cannot convince females to mate will not pass his genes along to any offspring, even if he is fast enough to outrun any fox and lives to a ripe old age. Darwin's realizations about these issues allowed him to develop the theory of sexual selection using the same logic he employed for natural selection.

Survival is secondary to sex, merely an adaptation to keep animals alive so they can have a shot in the sexual marketplace. The essence of sexual selection is that traits of beauty that enhance an animal's mating success will evolve even if they somewhat hinder survival, as long as they are not too burdensome, as long as the costs they impose on survival do not outweigh the benefits they deliver for sex. Although most species have about the same number of males and females, not everyone gets to mate. In many species, some males get more than their fair share of matings, while most males die as virgins. An individual's mating success is influenced by how sexually attractive he is perceived to be by potential mates. The peacock with the longer tail, the frog with a more variable call, and the fruit fly with sexier odors are more sexually attractive and chosen by more females as mates. As with traits for survival, when sexual beauty has a genetic basis, it is passed down from generation to generation as males evolve more seductive ornaments.

When Darwin put together his two great theories, natural selection and sexual selection, he went a long way toward explaining the diversity of life. Many unique traits evolve because they attract more mates. Of course, being attractive enough to be chosen by females is not the only way to enhance mating opportunities. Fighting off the competition is also effective. This book focuses on how sexual selection leads to the evolution of sexual beauty, but I should mention that sexual selection can also lead to the evolution of sexual weaponry to fight off competition for mating. This other side of the sexual selection coin has been covered in great detail by Doug Emlen in his book *Animal Weapons: The Evolution of Battle.*[9] But now let's travel to the cloud forests of Central America to return to the topic at hand, sexual beauty, and specifically to think about how the two sexes contribute to this phenomenon.

* * *

Consider what has been called the world's most beautiful bird. Birders from all over the world travel to the cloud forests of Central America to see the Resplendent Quetzal, or at least the male quetzal. The first time I saw one in the mountains of western Panama, my hands shook as I steadied the binoculars to peer at him through the fog in the forest's canopy. He had a light green body marked by a bright red chest, the blue iridescent patch on his head added to his collage of colors, and what made me shake was the sight of his sparkling two-foot tail. Perching above us in the towering forest, he seemed more like a Mexican piñata than a real animal. I also saw a female quetzal, but no matter. She lacked all of the male's fancy embellishments, and I hardly gave her a second look.

Although the difference in plumage between the male and female quetzal can hardly be more striking, the difference between them is more fundamental than their feather-deep beauty. It resides deep inside their bodies—in their gametes, cells that contain copies of all of the animal's DNA and can be fused with a mate's gametes to form new individuals and continue the cycle of life. The male's gametes, his sperm, are the smallest cells in his body, and there are lots of them. Meanwhile the female's gametes, her eggs, are the largest cells in her body, and there are fewer of them. This difference in gamete size defines the sexes, male and female, for all animals—everything else is secondary—even the external sex organs.

In humans and other animals, you can often correctly identify an individual's biological sex by the sex organs. Males, with small gametes, often have penises, and females, with large gametes, often have vaginas. But human sexual identity depends both on cultural and biological factors, such as brain development. An individual with female gametes, for example, could have a masculinized brain. In humans, there is a difference between sex and gender, the latter being a culturally created construct. Only humans have gender identities, a topic I will return to later.

But even for the rest of the animal kingdom, sex organs do not always correctly indicate an individual's sex, thus making the focus on gametes critical to determining biological sex.

One example in which the sex organs can give a misleading indication of sex occurs in some lice. Bark lice are small insects, about the size of a flea, often found scavenging algae and lichen under barks. Others, sometimes called book lice, feed on the paste that is used to bind books. A most bizarre group of species are less well known and cloistered away in some caves in Brazil, where they survive by feeding on bat guano. But it is not their diet that makes them so interesting. These females have a penis and, correspondingly, the males a vagina.[10]

The female bark lice use their penis as most penises were intended to be used, to insert into the opposite sex's vagina when mating. But unlike a typical male's penis, the female's penis does not deposit sperm. It telescopes to penetrate deep inside the male, where it then expands, anchoring the barbs on the penis to the male's vaginal wall, effecting a copulation that can last more than forty hours. The barbs provide such strong purchase inside the male that when a researcher tried to separate a mated pair, the male was torn in two. During the marathon copulation, the penis sucks up large volumes of sperm into the female's body, where they eventually reach and fertilize her eggs. Despite this role reversal in their sex organs, there is no confusion about their sex. By definition, the males are males because they have the smaller gametes, and the females are females because they have larger gametes. When it comes to sexual identification in nonhuman animals, it all comes down to sperm and eggs, and their difference in size is at the root of all the other differences between the sexes and the reason why there is sexual selection. To understand the evolution of sex differences, including sexual beauty, we need to understand why this difference in gamete size matters so much.

Let's unpack this idea of how gamete size is tied to the evolution of sexual beauty. The human egg has a volume one hundred thousand times greater than a sperm.[11] If your gametes are smaller, you can make more of them; a woman produces only about 450 mature eggs during her lifetime, while a man makes about 500 billion sperm during his. Since fertilization requires only one sperm and one egg, eggs are a limiting resource. In addition, once a female has her eggs fertilized, it can take weeks to months to get another batch ready. Males, on the other hand, can replenish their sperm supply within hours. In many species, once a female's eggs are fertilized, she is out of the mating game while she nurtures her inner embryo—a month for a guppy, nine months for a human, and almost two years for an elephant. While a female is tied up with her embryo, a male can go on mating. As with the sex-reversed bark lice, there are exceptions to the patterns in sexual selection. Male seahorses, for example, become pregnant, and a tropical male wading bird, the jacana, tends the nest while the female is mating with more males, "feathering" their nests with more eggs. But these examples tend to be not only exceptions to the general rule, but as we will discuss later, the exceptions that prove the general rule. And the general rule is that in most mating systems there is an excess of males ready to mate at any one point in time. This imbalance results in a sexual marketplace where many males compete for fewer females, a marketplace that features an abundance of courters and a limited number of choosers. All of this because sperm are smaller than eggs. So what can a male do to increase the chances of his sperm fertilizing her eggs? How can he compete in the sexual marketplace?

In some cases, males can control a resource that females want and need, which in turn makes a male more attractive. Males can control areas with food, nesting sites, and refugia from predators, all of which are important to a female with mating on her mind. Females can then shop and compare resources among males and mate with the most attractive choice. Of course, these resources are not free, as males have to fight for them, and sometimes rather fiercely. The weapons males use in these battles are varied and include larger size as well as assorted fangs, claws, horns, and antlers. The resources they defend can also vary, but all of them, in one way or another, are crucial to reproduction. For example, male damselflies defend areas of water with floating vegetation that females need to deposit their eggs; male fiddler crabs defend burrows that are used as refugia from predators as well as for mating; and men of the Kipsigis people in Kenya, as well as many other societies, accumulate wealth in various forms to recruit females for mating. And to the winner goes the spoils: the males with superior resources are more likely to be chosen to mate.

Although resource defense is one means by which an animal can enhance its sexual attractiveness, most of the interest in sexual selection centers on the beauty of the individual itself. The stunning male peacock is just the beginning. I have already discussed the quetzal's tail and the canary's song, and throughout this book we will look closely at an incredible diversity of traits that have evolved in the name of sexual beauty.

Thus far I've explained how natural selection and sexual selection came to exist as scientific theories, why sexual selection usually acts on males, and how sexual selection can result in the evolution of beauty. I have argued that to understand beauty we must understand the brain of those who behold beauty, but I have yet to illustrate how we can explore this relationship between beauty and the brain. Now I will focus on one species, the one that provided me an entrée into this field and led me to begin to explore the neural underpinnings of sexual aesthetics. This compelling example of sexual selection favoring the evolution of acoustic beauty comes from a frog most unassuming in his looks but quite audacious in his voice. In the next chapter, we will have a detailed look at how the sexual aesthetics of a female can drive the evolution of a distinctly beautiful, although somewhat dangerous, male voice. We will delve into her brain's function as well as its evolutionary history to uncover why she has judged this male voice to be so beautiful.

TWO

||||||||||||||||||||||||||||||||||||||||||||||||||||||||||||||||||||||||||||||||||||||||||||

Why All the Whining and Chucking?

Frog went a-courtin' and he did ride uh-huh . . . said "Miss Mousey
will you marry me?" uh-huh.   —*Old English folk song*

It was an orgy . . . of sorts. As is often the case, however, there were
not enough females to go around. So the males had to compete for
females, but this competition didn't involve brawn. There were no foot
races or arm wrestling matches or males pounding each other. This was
a beauty contest, and the beauty was in the males' voices. Males started
to sing, then their voices grew louder, and eventually their overtures to
the females grew richer with more notes. All of this for sex.

My own consuming interest in sexual beauty started on a piece of
land that connects the two great continents of the Western Hemisphere.
Until recently, at least in geological time, North and South America
were separated by a gap, where Panama is now but through which the
Atlantic and Pacific Oceans used to flow. This gap allowed beasts from

each ocean to freely mingle and interbreed, but it cut off interactions between the terrestrial organisms, allowing them to evolve in splendid isolation.[1] But this isolation didn't last. The Pacific and Caribbean tectonic plates moved toward one another; the continents collided; and about 3 million years ago, the Panama land bridge was formed. This union of North and South has been heralded as the most important geological event in the past 60 million years, the time elapsed since an asteroid crashed into the earth, causing massive worldwide extinctions. Cutting off the two oceans between the continents changed their flow of currents and brought about abrupt climate change. The most dramatic consequence of binding the North and South Americas was providing a conduit that allowed smaller mammals in North America to invade the south and then decimate South America's incredibly diverse mammalian fauna, including giant sloths the size of elephants.

The Panama land bridge also caused an inconvenience for humans. It made sailing between the two coasts of North America an arduous journey. To sail from New York to San Francisco, for example, ships had to round the tip of Tierra del Fuego—a mere six hundred miles from the Antarctic Peninsula. Our solution was to undo the Panama land bridge, to once again divide North from South, this time with a canal connecting the Atlantic and the Pacific.

The Panama Canal has a colorful history, part of which has made an important impact on tropical biology.[2] At the urging of Alexander von Humboldt, the canal was first planned in the mid-1800s by the Spanish government. The French began the project in 1881, but problems with construction and especially with numerous deaths caused by yellow fever and malaria caused them to cede the project to the United States in 1903. The Isthmus of Panama, however, was part of Colombia, and the Colombians were not willing to accede to the demands associated with a US-built canal in their own country. In what has become an oft-repeated US strategy in Latin America, President Teddy Roosevelt used some gunboat diplomacy to aid Panamanian rebels in achieving independence from Colombia. The Americans ultimately completed the canal a decade later and kept it for themselves. By opening the fifty miles across the Isthmus of Panama to ship traffic, the over-water voyage from New York to San Francisco decreased by more than half, from fourteen thousand to five thousand miles. The Panama Canal, now part of Panama, has recently been expanded to accommodate ships even larger than the one thousand–foot Panamax models that have trafficked through the canal during the past century.

What does this have to do with sexual beauty? The central part of the Panama Canal is Gatun Lake, which was formed when the Chagres River was dammed in 1913. Once flooded, hilltops in the area became islands. One of these—Barro Colorado Island (BCI)—became a nature reserve in 1923 and eventually the crown jewel of the newly formed Smithsonian Tropical Research Institute (STRI). Today, BCI is one of the most thoroughly studied tropical ecosystems in the world. Two of its best-known inhabitants are the túngara frog and the frog-eating bat.

In the summer of 1978, I boarded the Panama Railroad at the Balboa Station in Panama City, dripping with perspiration but restless and eager to finally reach BCI. In those days, the train was the preferred means of pedestrian transportation between the oceans. The main cargo was Panamanian businessmen and US armed forces, as the Panama Canal was still under US jurisdiction. The passenger list also included a band of disheveled and unkempt scientists, mostly graduate students in their twenties, and older but no tidier STRI researchers. We stuck out like sore thumbs compared with the immaculately groomed businessmen in their guayaberas and the meticulously uniformed US soldiers clutching their M16s.

The train stopped midway between the two coasts to disgorge us scientists at a stop called Frijoles. The station was merely a small cement bench and a tin roof that gave some protection from the intense sun or relentless rain, depending on the season. There was no sign of human settlement. Eventually, a small boat arrived to transport us to BCI. Seeing the island for the first time, I knew it would change my life forever, although I had no idea where this quest would take me and for how long it would consume me. Viewed from that boat, BCI could not have appeared a more peaceful, calming, and comforting sight.

From a distance, BCI looked like a poster of harmonious nature wrapped in a green curtain. Up close, bright splashes of yellow-flowered guayacan trees punctuated the flowing green canopy. As I first got to know the island, I saw a toucan calling from a treetop, bopping its oversized bill in rhythm with its call. Before this, I had only seen these birds on boxes of Froot Loops and advertisements for Guinness beer. A group of meter-long green iguanas dug their nests in the sand; bright blue morpho butterflies the size of my hand flitted along the trails. I soon realized that the ambience was even more pleasant at night, when more than thirty species of frogs gathered in choruses across the island to sing for sex.

All is not what it seems, however—first appearances can be deceiving. When I peered behind the green curtain to glimpse the evolutionary drama on BCI, what I saw was "nature red in tooth and claw." Parasites were everywhere: bot flies burrowed out of howler monkey flesh; grotesquely large tics dug into iguanas; malaria-carrying plasmodia swam through the blood streams of smaller lizards; and nematodes clogged the guts of frogs. Predators also abounded: birds caught and consumed those magnificent morphos on the wing; boas squeezed the life out of small, rabbitlike agoutis before dining on them; and large false vampire bats—the biggest carnivorous bat in the world, with a wingspan of a meter—swooped to pick up rodents scurrying across the forest floor and then devour them from head to tail, loudly crunching their bones. I was also introduced to bivouacs of army ants that dismembered any small animal in its path; acacia ants that fiercely attacked other herbivores that dared approach their defended food source; and the nocturnal cries of coatis, relatives of raccoons, which punctuated the night as females ganged up on males to expel them from their troops. BCI was not as peaceful as I first thought.

Maybe sex will be different I thought—perhaps it is more benevolent. It is a mutual endeavor for the benefit of both males and females, courters and choosers; at this point, I had hardly given the sexual conflict between males and females a second thought. My goal was to understand how members of one sex evolved traits to make themselves more attractive to members of the opposite sex, or how beauty evolves in the service of sex. My subject was a small, brown frog that haunts mud puddles, a frog that Panamanians called *túngara*. This plain-looking frog has a stunning voice that determines whether a female will find a male attractive enough that she will mate with him.

Túngara frogs had been studied for a short time in the 1960s by STRI scientist Stanley Rand, arguably one of the great tropical biologists of the twentieth century. Stan became my closest colleague, fellow traveler, and best friend for three decades until he passed away in 2005.

At that time, I wrote that second only to BCI itself, Stan was STRI's most valuable resource.[3] Few disagreed.

What I learned from the túngara frog launched me into a lifelong interest in sexual beauty. I have sought to understand how beauty resides in voices, in colors, and in odors, how it comes about in humans as well as other animals, and especially why it is we find some individuals so beautiful. Why do we and other animals have the sexual aesthetics that we do?

* * *

On my first day on BCI, I found evidence of the previous night's orgy. Small piles of white foam scattered about the pond's edge evidenced the debauchery that had played out the night before. I waited. Night descended quickly, as it does in the tropics, and the songs of insects and some chattering of night monkeys filled the dark as sheaths of moonlight penetrated the forest canopy. Then the chorus began.

Men and women use all sorts of ploys and accoutrements to attract sexual attention: deep voices, short skirts, tailored physiques, perfume and cologne, fast cars and expensive watches. When froggy goes a courtin', though, it is all about song. When the sun sets and their minds, or at least their brains and their hormones, turn to sex, male frogs become singing machines. A courting túngara frog might make more than five thousand calls in a single night.[4] There are about six thousand species of frogs. Most of them have loud and conspicuous mating calls, and each species has its own distinct sound. When I am in a chorus of frogs in Panama, the Amazon, Florida, or the East African plains, where there can be more than a dozen species calling at the same time, with a little preparation I can identify by ear each species without seeing a single one of them. What is all this singing about?

Males call to let females know who they are, where they are, and that they are ready to mate. These calls are designed not only to inform but to persuade, to charm, and to seduce. Males call loudly, incessantly, and add fancy notes to their calls, all traits that over evolutionary time have proven attractive to females. Once a female arrives at the chorus, she compares the calls of all these males and eventually decides which one is most pleasing to her ear, which male is most sexy. Her choice defines sexual beauty for this species. Choosing members of the opposite sex for mating based on sexual beauty is a common theme that plays out throughout the animal kingdom every time there is sex. Only the details vary.

The túngara frog is a small beast, only about thirty millimeters in length. Give males some standing water and they will call, be it from large ponds, small puddles, the overflow of streams, footprints of large mammals, the ditches around human settlements, and even in small aquaria in my laboratory. Anywhere from one to a dozen males might be calling in a square meter. Like the males of many frog species, male túngara frogs gather in a chorus to call at a breeding site once the sun goes down. The calls of these males seem to emanate from an old-fashioned video game. They begin with a *whine* that is about a third-of-a-second long and that can be produced by itself or followed by up to seven short, staccato sounds known as *chucks*. We refer to calls with only a whine as simple calls and those that contain a whine and chucks as complex calls. We will give these calls a closer look, or listen, in a bit.

The túngara frog chorus is actually more of a sexual marketplace than an orgy, and in that sense it is a female's paradise; the males are on display, and the females are the consumers. Males do little besides call, and they pretty much stay put while they are calling. A female enters the chorus when she is ready to mate, and I mean ready. If she does not mate in the next few hours, all of her eggs come oozing out of her; unfertilized, they are a wasted investment in reproduction and a failed attempt to get some genes into the next generation. She then has to wait another six weeks before she has a batch of eggs ready to be fertilized. But this rarely happens, as females are surrounded at the breeding site by a surfeit of males only too ready to mate. And she gets to pick.

A female gives her choice of a mate some serious thought. She will sit in front of one male for a time, often move on to others, and sometimes return to a male she has already sampled. She checks out the males by listening to what they have to say, that is, their whining and chucking. When a female decides to mate, she slowly moves to the male. He clasps her from the top. They are now mating, although the mechanics are a bit different from those to which we are accustomed.

Frogs don't have penises, but the males do deliver the sperm to the females, and fertilization takes place outside the body. In túngara frogs, the eggs are extruded into the water by the female while the male is clasping her from the top. He catches the eggs in his hind feet and spills some sperm on them. He then makes a nest for the eggs out of frog merengue. He does this by using his hind legs as a whisk, beating the eggs and the various fluids that come with both gametes to produce an exquisite foam nest. The foam nest keeps the eggs out of the water and away from aquatic egg predators, and it also keeps the eggs moist and allows them to survive short dry periods when the temporary pools of water they will soon call home might dry out. If all goes well, the eggs hatch into tadpoles in three days, and in another three weeks or so the tadpoles emerge as small froglets who will themselves either be charming or charmed in their own sexual marketplace.

The narrative of the túngara frog's sex life was uncovered during 186 consecutive nights of watching everything these frogs did from sundown to sunup—more than one thousand of them, all individually marked so I could tell them apart, record the males' voices, measure how often they mated, and figure out just what attracted females to a particular male. The short answer to the last question is that females always seemed to choose to mate with males who were making whines with chucks, and with males who were larger than average. But where did túngara frogs get their sexual aesthetics, why is it that the calls of some males are perceived as so much more attractive than others? The answer to this last question did not come easily.

If you think someone is beautiful, then he or she is beautiful—you are the decider. Sexual beauty emerges from the interaction of an individual's traits and the sensory systems and brains that perceive them. I find the *Mona Lisa* beautiful, and perhaps you don't. We both see the same arrangement of colors within the frame; we just process them differently. Remember, beauty is in the brain of the beholder. Why is it that female túngara frogs find the male's call, and especially the call with chucks, so attractive?

I could spend my life watching túngara frogs in puddles around Panama and learn little more about their preferences besides what seemed to be a strong predilection for chucks. How can we begin to glimpse what is going on in the female's sexual brain in order to get a better understanding of her sexual aesthetics, to uncover exactly what it is about a male's call that she finds so sexy? A few well-designed experiments can have the precision of a surgeon's scalpel as they enable us to look within the female and gain unparalleled insights into her standards of sexual beauty.

My team of túngara researchers collects female túngara frogs at breeding sites in Panama and brings them into our laboratory, where we place one of them (the frog, not a researcher) under a small funnel centered between two speakers in a walk-in acoustic chamber. We broadcast calls through each speaker, using the real calls of a male or calls we have synthesized electronically. In one of our first experiments, we gave females a choice between a whine broadcast from one speaker and a whine-chuck from the opposite speaker. Each speaker "calls" once every two seconds, and the two speakers alternate calls with one another. We close the chamber and observe the female from outside via an infrared camera. We remotely lift the funnel, and the female tells us which call she prefers by hopping to it. She has to travel about one meter to contact one of the speakers, which scaled to our size would be about an eighty-meter trek. The only reason a female frog approaches and contacts a calling male is to choose him as a mate. This simple experiment, called *phonotaxis* because the bioassay is movement toward sound, allows us to dissect the female's sexual aesthetics in great detail.

A male túngara frog's call does not need a chuck to attract a female. When placed in an acoustic chamber, a female will approach and make contact with a speaker that is broadcasting only a whine. A simple whine call is sufficient to attract a female, but it is hardly ideal in a competitive sexual marketplace. What about the chucks? A chuck doesn't do much on its own. If we play only a chuck from a speaker, something that never happens in nature, the females ignore the chuck. But the chuck is far from worthless, it just needs to be in the right context, and that context is the whine. If we put a simple and a complex call in head-to-head competition—a whine against a whine with one chuck—a female is five times more likely to go to the speaker broadcasting a whine-chuck than the one broadcasting a whine. The sexual potency of the chuck is extraordinary. Although it only increases the call's duration and energy by 10 percent, it increases the male's attractiveness by 500 percent. Give some thought to any change that we could make so cheaply in our own appearance that would have such an effect. If you can think of one, patent it!

Besides choosing males that make chucks, females are also more likely to choose larger males as mates. How does this happen in the dark of night, if she cannot see who is larger? The call seemed to be an obvious candidate: perhaps females can hear who is larger? In vocalizations of most animals there is a relationship between the frequency or pitch of the signal and the body size of the individual producing it. This is due to basic biophysics. Larger individuals have larger sound-producing organs—the larynx in frogs and mammals, file and scraper in crickets, syrinx in birds—and larger structures vibrate at lower frequencies. This is true in humans as well. Those deep, resonating voices of Sylvester Stallone and James Earl Jones are not produced by undersized men. Another aspect of our morphology that influences the structure of our sounds is the anatomy of the trachea above the larynx, which produces the formants of the voice. Women prefer lower-pitched voices of larger men,[5] and it has even been suggested that the descended larynx in humans evolved not to facilitate language, as has long been thought, but to lower the frequencies of vocalizations by increasing the length of the trachea above the larynx.[6] In fact, red deer actively lower the larynx when they roar to decrease the pitch of the roar and thus to increase their apparent body size. Túngara frogs have been doing some of their own evolving to make their calls deeper-pitched and more attractive. The larynx of the túngara frog is huge compared with that of other frogs of similar body size. In fact, the túngara frog's brain can fit inside the larynx. It seems that sexual selection has been favoring good looks, or at least good sounds, over big brains in this species.

Biophysics dictates that larger males have lower-pitched chucks. Do females prefer larger males because they are attracted to deeper chucks? This is another aspect of the female's sexual aesthetics that yielded quickly to experimentation. I created digitally synthesized calls that had identical whines but varied in the chuck's pitch. Over a large range of natural variation, females preferred deeper chucks. Although any advantage of women preferring baritone males is merely speculative at this point, I do know the advantage for this preference in túngara frogs: larger males fertilize more eggs. It is not that their sperm are better, but the greater fertilization success derives from a better mechanical match between the male and female. Remember, the male is on top of the female, as they mate "froggy" (not quite "doggy") style as they both release their gametes. If a male is too small, his sperm gets all over the female's back, and the stranded sperm are less likely to meet the eggs as they come out of the female.

It seems logical that this reproductive advantage to mating with large males is responsible for the female's preference for deep chucks, which, in turn, drove the evolution of the grotesquely large larynx in these frogs. But logical need not mean biological; it is a good start, however, to generate hypotheses about what really happened to give us the biology we are investigating. We will return to this issue shortly.

* * *

It is often said that the best science raises more questions than it answers. It should be added, however, that one of the most frustrating things about science is that when you finally get an answer, you often then have more questions than when you started. This is what happened to me. Our understanding of sexual selection by mate choice in túngara frogs raised a serious paradox. The chuck is a pretty small sound to have such a large impact on the male's beauty, and all males can produce chucks. Simple evolutionary logic predicts that males should spend the night chucking until they attract a mate. But this is not the case. Túngara frogs are reluctant to add chucks, and many prefer to just whine. But males should try to get as many mates as possible—they are males, after all, right?

As with humans, an individual's investment in being attractive is subject to social influence, as I will discuss at length in chapter 7. Men are more likely to flaunt their resources to a woman if there are other men around, and women flirt more with men when in a crowd of women.[7] There are two social settings that get túngara frogs to make more chucks. Males add chucks when other males call, and this escalates into a chorus of complex calls where the "whine-chuckers" prevail and "whiners" are in the minority. Females also seem to "flirt" with males to get them to make more chucks. If a male refuses to escalate from a whine to whine-chucks, a female sometimes gives him an actual body slam that, amazingly, he responds to by adding a chuck.

So again, why this reluctance to chuck? To understand why any trait evolves, it is necessary to understand not only the benefits it delivers but the costs it incurs. As in human economics, Darwinian economics predicts that it is the average benefit/cost ratio that determines the value of a trait and the degree to which it is favored by selection. Here, the currency is not euros, pesos, or dollars, but fitness—the number of offspring produced relative to others in the population. And unlike many economic transactions in humans, the goal is not maximizing relative benefits in the short term but rather over an individual's entire life.

Consider, for example, a cheetah hunting for food. I have seen them sprint across the savannas of East Africa in a blur as they ran down hares and gazelles. Running fast aids the cheetah in catching its food, and food is essential for survival—dead cheetahs don't mate. So cheetahs have evolved to run fast: they accelerate to 40 mph in three strides and reach a maximum speed of 70 or 75 mph, the fastest any animal can run.[8] Why not even faster? There are physiological constraints to running fast. The cheetah has a relatively large heart, and it pumps so rapidly as the animal races that a cheetah can only maintain its maximum speed for about six hundred meters. It is so overexerted by then that it flirts with brain damage, and when it does catch its prey, it has to rest before eating.

There is a big physiological cost of calling in túngara frogs. While calling, their metabolic rate, and thus the rate of energy consumption, increases by about 250 percent. But this cost does not explain the reluctance to chuck; adding a chuck to a whine is pretty close to free when it comes to energy. There is another cost of chucks, one that remained hidden to me for more than a year but that had been influencing the evolution of sexual beauty in túngara frogs for millennia: the cost of eavesdroppers.

Many of us have been embarrassed by things that we thought we were saying in private but that were actually overheard by others. We should never assume privacy. That has always been true for animals. When male túngara frogs call, females are not the only ones listening. One fierce eavesdropper is the frog-eating bat. Merlin Tuttle, who went on to found the world-renowned Bat Conservation International, was on BCI the year before I began work there, and he caught a bat, *Trachops cirrhosis*, or the fringe-lipped bat, with a túngara frog in its mouth.[9] As one of the world's experts on the ecology of bats, Merlin understood this was an odd food item for a bat and wondered what it meant for their lifestyle, where they lived, and in what habitats they foraged. He also wondered if the bats might hear the frogs' calls. He wanted to team up with me to explore this interaction between bats and frogs. When I received Merlin's handwritten letter (this was long before Al Gore "invented" the Internet) I was thrilled that this bat might hold the answer to the riddle of the túngara frog's chuck.

My excitement was quickly tempered, however, by what I knew about the biology of bats. I realized how unlikely it was that this bat could hear frog calls, let alone use them as cues to find the frogs. Bats are well known for their echolocation abilities.[10] They emit high-frequency pulses that bounce off objects, return to the bats, and are perceived as an acoustic image of the world around them. Their echolocation pulses are ultrasonic, "ultra" because they are above our hearing range of 20,000 hertz (Hz). The fringe-lipped bat's echolocation call is between 50,000 and 100,000 Hz. With their hearing range shifted to such high frequencies, bats were thought to be near deaf to frequencies that we can hear. Frog calls, on the other hand, usually have only low frequencies. The loudest frequency in the túngara frog's whine is only 700 Hz, and 2,200 Hz in the chuck, a bit higher but nowhere near being ultrasonic.

When Merlin, our assistant Cindy Taft, and I set up one of our nocturnal viewing stations near the Weir Pond on BCI, we saw these bats regularly catching and eating calling túngara frogs; on average, about six frogs per hour met their demise in the jaws of these bats. We also had spectacular photos of the bats in the act; Merlin is a magical photographer, and soon his photos of the frog-eating bat, along with a story of our discoveries, were gracing the pages of *National Geographic*. But were the bats listening to the frog's calls or using their echolocation to find them? We needed experiments.

How do you catch a bat? The easiest way is to a find the path the bats use to transit through the forest. Then, before it gets dark you string a thin "mist" net across the path. Bats can detect these nets with their ultrasonic calls, but only if they are paying attention. As with highway hypnosis, the bat's mind can wander and lose focus on its echolocation calls when it travels the same paths night after night. If something unusual is in their paths, they often fly into it. This is one of the ways that Donald Griffin discovered bat echolocation in the 1930s. In his lab, he had placed various objects in the bats' flight cage, which they deftly avoided using their skills in echolocation. Once the bats became used to where these objects were, he moved them around and the bats crashed into them. Griffin called it the *Andrea Doria* effect, after the famous ocean liner collision of 1956.

Although we had observed bats eating calling frogs, that did not mean that the bats were orienting to the calls. They could have just as easily been echolocating the frog's body. We gathered some circumstantial evidence in the field that frog-eating bats were attracted to the túngara frog's call. We put speakers broadcasting frog calls near the bottom of the mist nets in the forest; quite often the bats flew into the mist net right above the speaker. Since then, generations of researchers have used acoustic baits to catch these bats. More convincingly, we also placed pairs of speakers in the forest, one broadcasting the túngara frog's simple calls, whines only, and the other its complex calls, whines with chucks. The bats came out of the canopy, swooping right over the speakers. We did not know how many bats were involved, but of the more than two hundred passes by bats over the speakers, about 70 percent were to the complex call. The acid test, however, took place in a flight cage and mimicked the experiments I did with the female túngara frogs; we gave the bats a choice between a simple call and a complex call. Even though the bats and frogs are interested in the calls for different reasons—one for a meal, the other for a mate—they showed similar responses to the calls. Both frogs and bats are attracted to a simple whine, and both preferred the complex call to the simple one; about 90 percent of the bats' responses were to the complex call. Paradox resolved! Many years later, Rachel Page, now a STRI staff scientist, showed that bats could more easily locate calls with chucks compared with calls without chucks.

Adding chucks to the whine increases a male's success in attracting a mate, but it also increases his risk of becoming a meal. The males are at the tipping point between sex and survival: more chucks tilt the balance one way; fewer chucks, the other. A few years later, I worked with two neurobiologists, Volkmar Bruns and Hynek Burda, to show that frog-eating bats have adaptations in their inner ears that allow them to remain sensitive to the ultrasonics in their echolocation calls while extending their hearing down into the lower frequency range of frog calls.[11] As far as we know, none of the close relatives of the frog-eating bat share these neural adaptations for sonic hearing. If that is the case, then it seems that when frog-eating bats first encountered frogs in the rainforests of tropical America, they probably hunted frogs with their echolocation calls but were deaf to the calls of the frogs. But after some evolutionary tinkering of their ears, and undoubtedly some reorganization of the brain, they became the túngara frogs' greatest nemesis and put some checks on the evolution of sexual beauty in this species.

* * *

Preferences of any kind are a hard thing to nail down in humans or in other animals, and sexual preferences are among the hardest. A variety of studies have shown that women prefer men with more chiseled facial features,[12] peahens prefer peacocks with longer tails,[13] and female túngara frogs prefer calls with chucks. But where is the seat of these preferences, and what mechanisms regulate the preferences for one mate over another? What needs to change for a preference to evolve?

A behavioral preference is a phenomenon that results from an interaction between incoming stimuli and inherent biases in the sensory, neural, and cognitive processing of those stimuli. If we want to understand how preferences evolve and why they differ among species, we need to measure more than just the behavioral output. We need to understand what changes in the hardware contribute to these behavioral preferences.

To explain my desire to understand preferences at the level of the brain, let me give an analogy. We can compare the maximum speed of a cheetah and a leopard by measuring their velocity as they sprint across the savanna. The cheetah is faster, and it has evolved to be faster. But we really have learned little about how evolution has taken place until we go under the hood of this well-oiled running machine. If we also measure the biomechanical efficiency of the limbs, the size of the heart, and the contributions of aerobic and anaerobic metabolism to support such speed, we can say a lot more than just "cheetahs evolved to run fast"—we can actually describe what evolved.

We went under the hood of túngara frogs, and we found that both individual areas in the brain and complex networks that connect brain areas contribute to the sexual preferences for calls. We utilized two different approaches to determine how different parts of the brain respond to different sounds. In one approach, neurophysiology, electrodes record impulses from neurons in various parts of the frog's brain while different sounds are played. The electrodes record neural firing, and this allows us to determine which sounds elicit the most neural activity in different parts of the brain. In the second approach, gene expression, females are again exposed to different sounds. They are then sacrificed, and we slice the brain to identify the expression of certain genes that indicate neural activity has just taken place. These studies of the brain, combined with our detailed knowledge of behavioral preferences, have produced a relatively simple explanation for why females prefer certain calls.

As I will discuss in detail later, the most important mate choice decision an animal can make is getting a partner of the same species. If a female mates with a male of the wrong species, a heterospecific instead of a conspecific, she has usually wasted her substantial reproductive investment, as little of Darwinian value ever results from these mis-matings. In most species, courters exhibit characteristics that provide choosers with unambiguous information about their species identity.

As I said before, there are about six thousand species of frogs. Nearly all of them call, and the calls of all of those species are different. When females are tested behaviorally, they almost always prefer the call of their own males to those of other species. Where is the seat of this call preference? The entire neural circuitry of the auditory system, the decision-making system, and the behavioral output system biases the female to find these conspecific calls most attractive, most sexually beautiful. These neurons determine her sexual aesthetics, and these are the brain areas in frogs that must change when call preferences evolve.

In humans and frogs alike, hearing starts in the ear, the inner ear to be precise. The inner ear is a capsule in our head that contains organs of balance and hearing. Inside are hair cells embedded in membranes that discharge neural impulses when the membranes rock back and forth in response to sound or a change in the orientation of our head. The neurons that innervate the hair cells in the hearing organ form a large collection of neurons called the auditory nerve, which is the conduit of information from the inner ear to the brain.

As I will discuss in the next chapter, all sensory, perceptual, and even cognitive systems are nonlinear; that is, their neural and behavioral output cannot be predicted by only the stimulus input. This is a pretty obvious statement. For example, many potential stimulus inputs are not even detected. Of the entire electromagnetic spectrum, which extends from X-rays to radio waves, we are only able to see a small sliver of wavelengths, from 400 to 700 nanometers. We are blind to much of the ultraviolet that is seen by many birds and fishes. This restricted perceptual field occurs in all of our sensory modalities. Man's best friend, the dog, is sensitive to thousands of airborne odors, whereas our olfactory system is fairly anosmic compared with the dog's. The same is true of sound. Given our hearing range of 20–20,000 Hz, we are deaf to the ultrasonics that dominate most bats' auditory scene. Most bats, on the other hand, are deaf to most of what we have to say. The frog-eating bat, as we have seen, is an exception.

If we want to begin to understand how the túngara frog's brain codes the sexual aesthetics of sound, we have to first know what the frog's ear tells the frog's brain. Even within its range of hearing, any ear of any animal hears some frequencies better than others. We are just like other animals in this regard. Although we hear frequencies over three orders of magnitude, we are most sensitive to frequencies from about 2,000 to 5,000 Hz. Frogs, however, carry a restrictive hearing range to extremes. Instead of having one hearing organ in the inner ear, like all birds and mammals, it has two. One, the amphibian papilla (AP), is usually sensitive to sounds below 1,500 Hz and the other, the basilar papilla (BP), is sensitive to sounds above 1,500 Hz. Many years ago, Bob Capranica, while first at Bell Labs and then at Cornell University, showed that the tuning of these two inner ears is used as a pair of filters that match the more dominant frequencies of the mating call of its species.[14] My colleague Walt Wilczynski and I worked with Capranica to varying degrees and showed that the same is true in túngara frogs.

What does the ear hear? By playing the frog pure tones while recording neural discharges from auditory nerves, Walt showed that the tuning of the two inner ears is well matched to the frog calls. The AP is tuned to right around 700 Hz—a near perfect match to an average whine with a dominant frequency of about 700 Hz. The BP, as expected, is tuned to higher frequencies, 2,200 Hz to be exact. This is close to the average chuck's dominant frequency of 2,500 Hz but favors frequencies that are lower than average. One reason the túngara frogs prefer conspecific calls is that these contain sounds that they hear better, and one reason they prefer lower-frequency chucks of larger males is that those calls are closer to the peak tuning of the BP than the average chuck. This means that the females perceive calls of larger males as louder than calls of smaller males. Thus our first venture under the hood showed that the tuning of one of the frog's inner ear organs, the BP, helps to define one part of her sexual aesthetics, namely, her preference for the lower-frequency chucks of large males.

The ears serve as the conduit to the brain for all of the neural stimulation derived from sound. The brain is where the real action is when it comes to preferences of any kind. To describe this neural machinery, we supplemented our neurophysiology with the second approach to decoding these preferences. Kim Hoke worked with me and Walt to use gene expression to visualize the location and quantity of neural responses the túngara frogs showed to different types of sounds.[15] After a túngara frog was exposed to one type of sound, either the whine, the whine-chuck, white noise, or the call of a heterospecific, she was sacrificed; her brain was thinly sliced; and then a molecular probe identified where there was expression of particular genes that are indicators of neural activity. In this way we were able to visualize how neural activity throughout the entire brain varies in response to these different sounds.

It would be ideal for us scientists if there were a single neuron that encoded sexual aesthetics; "push her button" with a sexy sight, sound, or odor, and bingo, her sex neurons fire and she is turned on, in love or lust depending on the species. But these types of single cells or feature detectors now seem to be the exception rather than the rule. Decisions, such as what to eat, where to sleep, and with whom to mate, are more likely to result from the summed responses of entire populations of neurons. This makes sense given that most sexual stimuli are a complex mix of stimulus variables, such as the duration, frequency, and amplitude of sound components, and different neurons are tuned for different types of stimulations.

This is what Kim found to be true in túngara frogs. The hindbrain of the frog harbors a large area or "nucleus" responsible for auditory analysis. Kim showed that across neurons in the auditory nucleus there is more neural firing in response first to the whine-chuck and then to the whine compared with other kinds of sound. Not only do the sexually attractive signals elicit more neural stimulation in the auditory area of the brain, but they also change the activity relationships among other brain regions; that is, the degree to which activity in one brain region is correlated to activity in another brain region, also known as the *functional connectivity*. When a túngara frog hears a túngara frog call, compared with calls of other species, there is an increase in correlated neural activity in the neural circuits that generate decisions, in the circuits that are involved in the sensation of reward, and in those that generate motor output, namely, phonotaxis to her mate. Although some details are lacking, we now have a basic understanding of the underlying sensory, neural, and cognitive processes that generate the sexual aesthetics of the female túngara frog. The seats of these preferences are no longer a mystery; we can open the brain and point right to them.

* * *

Biology raises questions about how things work and also why they evolved to work that way. We usually refer to these two sets of questions as proximate and ultimate ones, and most studies of biology remain firmly entrenched in one of those domains. But not this study. The túngara frog has become one of the best-known "model systems" in sexual selection because my colleagues and I have been able to answer questions in each domain and, even more important, because we can use information from one domain to explore questions in the other.

Earlier in this chapter, I pointed out that female túngara frogs prefer the lower-frequency chucks of larger males and that these larger males fertilize more of her eggs than do smaller males. And just above you read that this preference for larger males results from the female's ear, specifically her BP, being tuned to the frequencies that are slightly lower than the average chuck. Thus females receive more neural stimulation in the BP from larger males compared with smaller ones. Evolutionary logic would suggest that the tuning of the female's ear evolved to be what it is because it generates a preference for larger males, who fertilize more eggs. Thus females whose BPs are tuned to lower-frequency calls of larger males will have a selective advantage over females with BPs tuned to higher-frequency calls of smaller males. The evolution of this particular sexual aesthetic, preference for lower-frequency chucks of larger males, evolved because it boosted the female's Darwinian fitness. This is logical, but as it turns out, it is not biological; this is not what happened. How do we know this?

The túngara frog has about eight close relatives, all of which are in South America. Half of them are in the Amazon, on the eastern side of the Andes, and the other half are on the western side of the Andes. Stan Rand and I made numerous trips to record calls and collect individuals of these eight species. We worked in every country in Central America, from Mexico to Panama, in the Andes of Peru and Ecuador, in the Amazon basin in Ecuador and Brazil, and in the Llanos grasslands of Venezuela. With the exception of some populations of an Amazonian species, all of the other frogs in this group have males that produce whines but not chucks. What is the tuning of their ears? We brought these frogs back to Walt Wilczynski, who characterized it, as he had previously done for the túngara frogs. These species were remarkably similar.[16]

In each of the species, the tuning of the AP matched the frequencies in their whines. Since most of them did not make chucks or other higher-frequency notes, we knew they did not use their BPs in communication. But they all still had a BP, and those BPs were tuned. Amazingly, they were all tuned to the same frequencies to which túngara frogs were tuned. For the túngara frogs this makes sense because the BP tuning matches the chuck, but most of these other frog species were chuck-less.

Evolutionary biologists rely on a principle called *parsimony* to interpret how events happened in the past. Parsimony posits that, all else being equal, the simplest explanation is more likely to be the correct one. Think of the human heart. We have a wonderfully adapted four-chambered heart that excels at oxygenating blood. But so do all of the other 5,500 species of mammals in the world. Did this exquisite adaptation evolve 5,500 times, once in each species, or did it evolve once in the ancestral mammal and was then inherited by other mammalian species as they arose? The answer is plainly the latter.

When we apply this same logic to the túngara frog and its close relatives we conclude that the tuning of the BP, which they all share, did not evolve in each species separately but is shared through a common ancestor. That means the BP tuning existed before chucks. This totally flips our thinking about how the preferences for lower-frequency chucks evolved. Females did not evolve their tuning because of the benefits of preferring larger males, but instead, when males evolved chucks, they evolved frequencies that matched the preexisting tuning of the female's ear. We called this process *sensory exploitation*,[17] and along with a more general process I discuss in the next chapter, *sensory drive*, this idea caused an intellectual revolution, or what the philosopher Thomas Kuhn called a *paradigm shift*, in the field of sexual selection.

I hope that this stroll through the sex life of a simple tropical frog convinces you that beauty and the brain are inextricably linked. It convinced me that the brain is the missing link in our understanding of sexual selection for sexual beauty not only in túngara frogs but throughout much of the animal kingdom.

THREE

||||||||||||||||||||||||||||||||||||||||||||||||||||||||||||||||||||||||||||||||||||||||||

Beauty and the Brain

Beauty in things exists merely in the mind which contemplates them.
—David Hume

THERE IS NO BEAUTY WITHOUT A BRAIN . . . just as there is no sound when a tree falls in the forest if there is no ear to hear it. Beauty is not only in the eyes of the beholders but in their ears, noses, taste buds, and touch receptors. These are the sense organs that are first stimulated by the world around us and that channel the outside world into our inside brain, the final destination, where our percepts of nature are formed and where our sexual aesthetics emerge. To understand beauty we need to understand the brain, and to understand sexual beauty we need to understand the sexual brain. The sexual brain is not a discrete module, but as we will see, it involves all the parts of a neural system that analyzes and makes decisions about sex, as well as modulates the way one feels about it. What makes the sexual brain complex and sexual aesthetics so unpredictable is that most of these neural processes are multipurpose, shared across domains, and recruited to perform different tasks. For example, animals use the same retinas and photopigments to assess both food and mates. And our affective reactions to sex, drugs, and rock and roll are modulated by the same reward centers.

In the next three chapters we will explore beauty through each of three major senses: visual, auditory, and olfactory. In this chapter, we will consider how and why the brain has a sexual aesthetic, why it varies among different animals, and how the other domains in which the brain must function influence what we decide is beautiful.

* * *

We all know people who seem to be in their own world. This is true for animals as well, and the German cyberneticist Jakob von Uexküll even coined a word for it—the *Umwelt*.[1] Different animals, he argued, can live in the same physical location but inhabit different sensory worlds, so much so they might as well be living on different planets. *Umwelten* differ among species as much as their morphology and their DNA. Understanding how animals perceive their environments, and how individuals of the same species can differ in their perceptions of the world, helps us begin to fathom how they experience beauty.

For eons before the Irish author Bram Stoker wrote *Dracula*, bats have frightened the masses and confounded the naturalists.[2] Bats are the only mammals capable of self-powered flight; they have small eyes, and yet they fly around on the darkest of nights able to navigate as if they had a supernatural sense. In the late 1700s, the Italian Catholic priest and scientist Lazaro Spallanzini tortured bats as he searched for their sixth sense.[3] He burned out their eyes and filled their ears and noses with wax in an attempt to decipher just how they could find their way around in total darkness, but he never quite figured it out. It was more than a century and a half later, in the 1930s, that Donald Griffin and Robert Galambos solved the puzzle in a more benign way, using new technologies to record ultrasonic sounds and noninvasive behavioral experiments to demonstrate the ability of bats to navigate in the dark.[4] They provided the first real insights into the bat's world of ultrasonic echolocation. The returning echoes of these calls, which are out of our hearing range, provide bats with an acoustic image of their world somewhat similar to our visual image. But only somewhat. The famed philosopher Thomas Nagel asked rhetorically, "What is it like to be a bat?"[5] Scientists might one day be able to describe in great detail all the behavioral and neural mechanisms that enable echolocation, but, Nagel argued, we will never know what it is like to be a bat because we can never have the same shared sensory and conscious experiences.

Beauty is also a sensory experience, and we describe it as such: a painting looks beautiful, a meal smells delicious, and a song sounds enchanting. Different animals experience sexual beauty in different senses; moths, fishes, and mammals are very keen about smelling one another, while crickets, frogs, and birds spend a lot of time listening. If we want to understand the diversity of sexual aesthetics that gives rise to the diversity of sexual beauty in the animal kingdom, we must understand how sexual beauty emerges from the senses and the brain. Extrapolating from Nagel, when we catch a whiff of a buck's musk during his rut, we might not share the same ecstasy as a doe, but if we probe her olfactory system, we can at least understand why she is in ecstasy.

To understand how animals perceive beauty we will begin at the sense organs—the ears, eyes, and noses—as they are the portals that connect the individual to its outside world, the conduits through which sensations flow to the brain. These sense organs are gatekeepers, and not all sensations can enter.

I have never seen more spectacular rainbows than those over the Dingle Peninsula on the west coast of Ireland. They rise out of the sea, arch overhead, and then plunge into the green coastal hills. It is hard to believe there is not a pot of gold at the end. A rainbow is refracted sunlight sorted into colored bands by particles in the air. We see the longer wavelengths as red bands on one border of the rainbow and the shorter wavelengths as blue bands on the other, with greens, yellows, and oranges in between. I doubt that a seagull flying overhead feels what I feel when I see a rainbow, and I am certain it doesn't see what I see. Gulls, and many other birds, have vision that is shifted to shorter wavelengths; they can see ultraviolet (UV) light, which we can only sense when our skin is burned by UV sunlight. Thus a seagull looking at a rainbow in Ireland should see extra bands of color beyond the blues that are visible to me. Bees can also see into the UV range of light. This is why flowers that attract bees to pollinate them often decorate their petals with a seductive tapestry of UV-reflecting markings that point to the flower's sex organs, seemingly urging the bees to "come see my etchings." Similar differences among animals occur in other senses. We can hear a bat's wing as it flutters past us in the night, but we are deaf to the bat's ultra-sounds echoing all around us. We can hear an elephant trumpeting, but not the infrasounds it uses to communicate with other elephants miles away. A large range of odors are even further out of our reach. Not even Nagel would contemplate what it would be like to have a dog's sense of smell; it would be too mind-boggling. Since we can only appreciate what we sense, differences in sense organs generate differences in sexual aesthetics among animals. This is the primary reason why sexual beauty takes so many forms.

But shouldn't all sensory systems provide an accurate description of the world if, as all evolutionary biologists would argue, they evolved to facilitate survival and reproduction? Would it not be best to have a full and unbiased estimate of the world around us, rather than one limited to certain bands of excitation? This is another example of an assertion that seems logical but is not biological; it just doesn't happen. Each of our sense organs responds to only a sliver of the world in its modality. As I just noted, human ears are sensitive from 20 to 20,000 hertz (Hz), making us deaf to infrasound (< 20 Hz) and ultrasound (> 20,000 Hz). Our eyes, by definition, are only sensitive to visible light with wavelengths from 400 to 700 nanometers (nm), an astonishingly narrow band of the electromagnetic spectrum, which extends from 0.01 nm gamma rays to radio waves longer than 1,000 nm. Our olfactory system, as well, seems anosmic in terms of the bouquet of volatile compounds in the environment that could be sampled, and compared with those of other animals who sample this world much more generously. And even within the accessible ranges of stimuli, our hearing, vision, and olfaction are all "tuned" to be more sensitive to a subsample of the stimuli that are sensed.

Why are our senses so stingy? There are two main reasons, constraints and adaptations. We just don't have the equipment we need to access all of the world around us. Short, ultraviolet wavelengths pack a dangerous amount of energy that can wreak havoc inside our retinas, while infrared wavelengths have too little energy to be captured by our photoreceptors. There are also trade-offs. Designing an ear sensitive to ultrasonic fre-quencies is usually done at the expense of hearing low frequencies.

There are also adaptive reasons for our restricted sensory worlds. In today's realm of "big data" we know that obtaining information is often the easy part. Computationally processing the data into meaningful patterns remains the primary challenge. Sequencing a genome is now a piece of cake; figuring out what it means is a very different story. Brains have the same problem: computation is expensive, and the more information pouring into our brain, the less efficiently we process it. Sensory channels are one way to filter out the noise from the signal before it gets to the brain. The only sensory experiences that matter in an evolutionary sense are the ones that increase or decrease our ability to survive and reproduce. Within the range of stimuli that a biological entity could possibly access, we expect our sensory organs to be most sensitive to what matters most. This is the case in the hearing of many frogs; their inner ear organs are tuned to best hear their species' calls—what sound could be more adaptive than that! You might suggest that it would be equally adaptive to have their hearing also tuned to the echolocation calls of the frog-eating bat. This might be a case where adaptation meets constraint. We assume, but certainly do not know for sure, that there are some design constraints that would prohibit the ear of the túngara frog from hearing sounds in the ultrasonics in addition to being sensitive to the much lower frequencies of their whine and their chuck. On the other hand, there is a species of frog in China that calls in the ultrasonic, and seems able to hear these calls.[6] This further suggests that túngara frogs should be able to detect bat echolocation calls, but they don't. At some point we will probably know why.

* * *

Our sense organs introduce the first biases in what we perceive, and they establish the foundation of our sexual aesthetics. But the yeoman's work takes place in the brain. Each sensory organ sends information to our central processing center, where it is passed from one processing hub, or nucleus, to the next, each more finely sculpting a percept from what we see, hear, and smell. In crickets and frogs, neurons in the brain are tuned to different properties of the species' mating calls, such as call rate, pitch, and duration. These neurons can combine their information such that the right call matches the percept of an attractive mate while other sounds, calls of other species, and even some other males of the same species fall short of meeting these criteria. As noted in the previous chapter, when a female túngara frog hears a whine, or better yet a whine-chuck, the main auditory center of her brain sparks with enhanced excitement compared with when she hears other sounds blowing in the wind or the whining of a different species. The same happens with sexual odors in fruit flies. Although the trip from the sensory channel to the brain is more direct in flies, groups of neurons in the brain are exquisitely sensitive to sexual odors and unresponsive to other kinds of smells.

Male fruit flies have a sexual pheromone, cVA, which stimulates courtship in females and suppresses it in males.[7] The molecular structure of this pheromone is complementary to the structure of one specific olfactory receptor in the fly's antennae called Or67d. The cVA molecule fits snuggly into Or67d, like a round peg in a round hole; this fit is the signal that the female is sensing a male of her own species. Square pegs, such as the pheromones of other species, don't fit and thus are not identified as an appropriate mate. When this perfect fit takes place, Or67d sends a message to the fly's brain, where it combines with other incoming stimuli to trigger the female's sexual attraction to the male—in response she courts him. This is not the only stimulus important in fly mate attraction, but stimulating this receptor is both necessary and sufficient to get females to court. How important is this receptor in defining the female fruit fly's sexual aesthetics? When researchers replace this receptor with a moth pheromone receptor, the mutant flies will then court when they smell a male moth!

Even for animals who seem focused on one sensory modality, the world is a multimodal place. Yes, we do tend to think of courtship in different animals in the context of one sensory modality or the other. Fish and butterflies are often visual courters, moths and mammals use odors, and frogs and crickets rely on sound. Although many animals court primarily in one modality, most are multimodal because choosers seem intent on extracting as much information as they can from a courter's display. Let's think about people speaking. Our lips move, which allows the sounds to escape from our mouth, and as this happens, our lips shape the sound into discernible phonemes. But the lips incidentally provide information about what we are saying, enough information to help us understand speech in a noisy world and for a lip-reader to understand speech in a silent one. The frog vocal sac is similar to lips in that it is needed by the caller to make sound, but it is also integrated into the listener's perception of the courter. In a number of frog species, females are more attracted to calls when they can see the male's vocal sac pulsing along with it.[8] This visual cue, however, doesn't do much for females if it is not tied to a call, just as the swinging arm of a metronome does little for our sense of rhythm without the sound of the beat.

Even flies have multimodal sexual aesthetics. The cocktail of modalities used in their courtship and mate choice includes sounds or "love songs" from their vibrating wings, visually detected athletic dances, the "taste" of a potential mate, and as noted above, sexual fragrances. All of these different types of stimuli are integrated in the brain and emerge as defining an attractive mate in a very specific manner. For example, there is a lot of groping between fruit flies when they court, and since they have taste receptors covering much of their bodies, this means a lot of tasting goes on during courtship. Getting a taste of the right female activates the male's taste receptor (more romantically known as $ppk23+$), which then enhances his attraction to the female he just tasted.[9] But she only is perceived as more attractive if the taste is combined with the right olfactory and visual stimulants—cVA being one of them. For all animals, and especially for us, one of the important tasks of the sexual brain is to bring together the stimulation from the different senses, integrate them, and then determine how this emergent concept of a potential mate matches up to our sexual aesthetics of a beautiful one.

Each sense is biased toward the sexual traits that stimulate it, and the brain to what combinations of stimuli it deems sexually attractive. But what are some of the rules that dictate these sexual aesthetics? As I suggested above, the brain has evolved to detect things that matter. What can matter more than getting the right mate? By the "right mate" I don't mean the best mate; we'll get to that later, but first things first. And the first thing is to find a partner who works, a partner whose gametes will fuse with yours to produce viable offspring. This first and most important criterion for a mate is that it be the right species, and the right species is your species, whomever you might be.

Choosing the right mate, the right species, is a crucial component of any animal's sexual aesthetics because genes do not act alone but are part of an integrated genome, and genes have evolved to be functional in their particular genetic background, with other genes of the same species. Different species of platyfish, for example, have different tumor suppression genes.[10] When different species are crossed, the functionality of these genes is disrupted, and the hybrids are cursed with melanoma, the same skin cancer that kills more than fifty thousand of us each year. In general, cross-species matings are a bad thing because genes of different species are often not compatible. Owing to various kinds of genetic incompatibilities, fertilization usually does not take place, and if it does, development often goes awry. The cost of this mate-preference error is a big one, especially for females, whose large investment in eggs is wasted. Luckily, the brain is pretty good at figuring out the correct species, because it makes choosers more attracted to traits of their own species (conspecifics) than to those of different species (heterospecifics). The details of how the brain accomplishes this mating match vary, but the general principles are similar among species and across sensory modalities.

In the previous chapter, I noted that all of the six thousand species of frogs have a unique species-specific mating call. These sounds provide enough information for the female's brain to correctly identify males of her species, as opposed to males of other species, as long as it is wired correctly. And the brain is wired to do this for every species that has been examined. The auditory system of frogs is biased to being more responsive to combinations of sounds from their own species. This is accomplished, as noted above, by sets of neurons that are biased to the particular patterns of sound that are peculiar to her species. The same is true for the auditory systems of crickets and birds, the visual systems of fishes and butterflies, and the olfactory systems of moths and mammals. The first attribute of an animal's sexual aesthetics is usually based on getting a mate of its own species. This is why sexual attraction rarely extends to other species.

This first rule of sexual attraction can have some important unintended consequences. Within a group of male sage grouse at a lek or male fish displaying on a coral reef, some individuals might better fit the profile of "conspecific" than others; they all are "right" mates, but some seem more right than others, and these will be preferred as mates because they better match the brain's sexual aesthetics. There might be nothing else better about the preferred mate except that she or he is more sexually beautiful, but no more healthy, wealthy, or wise than the others.

Once choosers have identified a group of conspecific mates, we expect them to move on to their next criterion and identify "good" mates. As we will see in the next chapters, certain attributes of sexual traits might indicate qualities of a courter that are beneficial to a chooser. If large antlers, broad shoulders, or a rich song indicate that a potential mate has control of better resources, will be a better parent, or has genes that are more compatible, then we would expect the brain to evolve to add these traits to the chooser's sexual aesthetics. The second rule of an animal's sexual aesthetic, therefore, is to not only get a conspecific mate (rule 1) but to get a good one (rule 2). Ultimately, a good mate is defined as a mate that will increase the chooser's number of offspring, or in other words, one that will increase the chooser's Darwinian fitness.

So far we have been concentrating on how the brain is wired to deliver conspecific and good quality mates by linking the perception of sexual beauty to these types of individuals. If we think of how animals evolve to survive in their environment, we usually envision the environment as an agent of selection that causes the animals to evolve traits, the targets of selection, that adapt them to the environment. Colder temperatures (agent) select for thicker fur (target) in many mammals, but this does not result in the environment now evolving even lower temperatures. The sexual brain, however, is both a *target* of selection, because it evolves in response to selection to deliver certain types of mates, and also an *agent* of selection, as it drives the evolution of sexual beauty in the opposite sex. Thus one way in which sexual beauty evolves is for courters to develop traits that exploit the chooser's sexual aesthetics. This is what happened in the túngara frogs as the males evolved a chuck to exploit the preexisting tuning of their BP hearing organ. This is what we call *sensory exploitation*.

I will review many examples of sensory exploitation in the next three chapters. A most famous one was revealed by studies of platyfish and swordtails by Alexandra Basolo when she was a graduate student.[11] Both types of fishes are common in your local pet store. They are close relatives to one another but differ in the presence or absence of one particular sexual trait. Male swordtails have a long sword extending from the tail. Female swordtails tend to prefer males who have longer swords. What about platyfish? They lack swords, but what if a male were suddenly to evolve one? Basolo answered that question by attaching a plastic sword to males and giving females a choice between a normal male platyfish and the artificially endowed ones. The female platyfish preferred sworded males. This means that female platyfish have a hidden preference for swords even though their males do not have swords. Since the platyfishes and swordtails are each other's closest relatives, this means that they share a recent common ancestor. Basolo applied the evolutionary principle of parsimony to this problem, much as I applied it to túngara frog hearing and túngara frog chucks in chapter 2, and concluded that platyfish and swordtails share a preference for swords inherited from their recent common ancestor. So when she added a sword to male platyfish, there was an immediate preference for it. Basolo didn't have to hang around the lab until the female platyfish happened to evolve this preference, nor would a male platyfish have to wait around to be considered more beautiful if he happened to evolve a real sword. He would be an immediate hit with his females.

Hidden preferences are often lurking in animals' sexual aesthetics, masked to others because there are yet no sexual traits to elicit it. But if that trait evolves, one that matches or exploits this particular sexual aesthetic, then it is immediately deemed to be sexually beautiful, and, all else being equal, it should soon evolve to be common among males. This notion of how sexual beauty evolves was virtually unknown before 1990 until a few other researchers and I developed this theory. Now it is thought to be one of the major factors driving the evolution of sexual beauty.

It would seem that sworded swordtails and chucking túngara frogs are in the vanguard of evolutionary development among similar species, as they and they alone have recently acquired these sexually desirable traits. An alternative, however, is that they are the last species to retain the trait, which has died out owing to competing pressures, such as predation risk. We must remember that what evolution giveth, evolution can taketh away.

* * *

When males "exploit" a female's sensory system by evolving a trait that attracts her to mate, the exploitation does not mean that choosing these males is costly to the females. When a male's signal better matches the female's neural biases, it is usually more easily perceived, and thus the female's mate choice is made more quickly and efficiently. Breeding areas are usually dangerous places because of all of the eavesdropping predators and parasites, so choosing fast often means choosing safe—fast sex is safe sex. Thus female choosers often benefit when their senses are "exploited" by male traits that evolve through sensory exploitation.

Sometimes sensory exploitation does lead to negative consequences for choosers. Yet even when there is an apparent cost to being sexually exploited, the true cost must be calculated within the context of all of the chooser's other activities, which can sometimes reveal that the price is not as costly as we might think.

A good example comes from plant-animal sex. Most flowers reward their pollinators with nectar or pollen for assisting in their reproduction.[12] Deceptive orchids are a big exception. These orchids use sensory exploitation by mimicking the silhouettes and odors of female bees to attract male bees that have evolved sensory systems and neural pathways that respond to these shapes and smells when searching for mates. The male bee attempts to copulate with one of these flowers and in doing so picks up some pollen. Eventually, the bee realizes the object of his desires is not a female, and he flies off. The next orchid that dupes him is pollinated by the pollen he carries, and he picks up more pollen from this second object of his misplaced sexual attention, which will be delivered to the next deceptive orchid he visits. He will finally find a female, but they are hard to come by. This waste of his time appears costly at first glance.

How stupid can this bee be? To be fair, we have to consider his attraction to mates as a signal detection problem. Consider that a set of stimuli can indicate either that another animal (or a plant mimicking one) is an appropriate mate or is not. An animal can make two types of correct decisions in this circumstance: accept appropriate mates and reject inappropriate ones. But we all make mistakes, and in signal detection theory there are two of them: a false alarm (incorrectly identifying the inappropriate individual as a mate) or a miss (incorrectly rejecting an appropriate mate). Which mistake would you rather make? It depends on the costs of each mistake. If mates are plentiful and the cost of copulating with a flower is dear, then it would be better to have a high threshold for mate recognition even if this would result in missing out on some real matings. But if mates are hard to come by, as in the case of orchid bees, and it is not very costly to ravage a flower, then it is best to lower your acceptance threshold. When you finally encounter a female, you do not want to pass her up, and so what if you have to fool around with some flowers along the way? When we first consider the decision of a bee to be attracted to an orchid it seems pretty maladaptive, but when we consider the decision in the broader context of the bee's life, it makes perfect sense. The small cost of ravaging a flower is less than the large cost of rejecting a real female as a result of having too picky a percept of sexual beauty.

Finally, if mating with an exploitative courter does result in a large cost to the chooser, then we would expect that the chooser would evolve a new response, perhaps a lack of response, to the exploitative trait. That is, the sensory biases that leads to sensory exploitation would change over evolutionary time. One way this could happen is by females evolving to abandon their biased preferences in favor of new preferences for sexual traits that are reliable indicators of important qualities of courters, those qualities that actually impact the number of successful offspring.

We have already seen how the importance of mating with the right species can influence the evolution of preferences for traits that indicate a courter is a conspecific. There are also traits within a species that might indicate an individual is a "good" mate, and we might expect choosers to evolve a preference for these traits when making mate choice decisions. Amotz Zahavi, a renowned sociobiologist, suggested that courters who exhibit costly traits, such as the peacock's tail, are demonstrating to choosers their underlying physical vigor in being able to bear this handicap.[13] If there were a genetic basis for the courter's ability to survive with this handicap, then he would pass down these genes to his and his partner's offspring, and tail length would be a reliable indicator to females looking for a "good" mate. We will slog through some of the nuances of the handicap principle in chapter 8.

* * *

The brain might be our most important sex organ, but it also has other things on its mind. The brain and its sensory systems are crucial for detecting all aspects of our environmental and social worlds, making sense of all those incoming data, and then responding in an appropriate manner. The brain has priorities, and in some instances it has to optimize how it performs tasks in one domain, which then influences how it performs in other domains. Food and sex are a good example of this interaction.

Food and sex have been intertwined in our culture for ages. Dinner is often treated as a part of a courtship ritual; oysters and chocolates are thought to have aphrodisiac properties; and the cherry is a symbol of virginity. We even refer to our sexual desires as appetites. The association between food and sex is a bit different in nonhuman animals, and the interaction less transparent. In some cases, the properties of what animals see best can be explained by how the eyes evolved to find food. In primates, for example, color vision is thought to have evolved to find colorful ripe red fruit. The reactions of some male monkeys to the bright red hindquarters of receptive females, and our own attention to red and green traffic lights and the colorful splashes on a Jackson Pollock painting, would not be possible if our ancestors only searched for food that varied in the grayscale. In other species, especially in fishes, evolution tinkers with the wavelengths to which the retina's photopigments are most sensitive in order to detect food in the complex ambient light that characterizes life underwater. The preferences of some fish for the bright courtship colors of males developed only after the eyes evolved biases to those colors for foraging tasks.

In the next three chapters I will discuss how sexual beauty is sensed by our visual, auditory, and olfactory systems, and in each case I will illustrate how these biases often first evolved to deal with needs other than sex. This general influence of the environment on the evolution of our senses is called *sensory drive* and is an important process that sets the stage for sexual aesthetics throughout much of the animal kingdom.

* * *

Sensory drive can be applied to higher-order brain functions that are not specific to any one sensory modality but instead are general processors that operate in any or all of our senses and can be applied to many domains. These are cognitive processes, and like the sensory processes just discussed, they too can have an important influence on an individual's sexual aesthetics, even when they evolve outside of the context of sex. I will review three of these processes now to set the stage for later analysis. So let us begin to get a sense for how habituation, generalization, and laws of comparison can be an important part of sexual aesthetics.

Nate Silver, the famed baseball sabermetrician and political pollster, asks a fundamental question in his book *The Signal and the Noise*: How do we tell them apart?[14] Animals are pretty good at this; they just ignore the noise and attend to the signal. Ideally, sense organs filter out a lot of noise from a stimulus before it even reaches the brain, but for the noise that gets through, the brain has an excellent way to deal with it: habituation.

It makes little sense for a shore bird to listen to and process the constant noise from the waves breaking on the shore; it is better off to use its brain to focus on searching for small fish in the surf and listening for the screech of hawks from above. Who has time for noise, especially since it is always pretty much the same? Usually it is best to ignore it, or more specifically, habituate to it. Besides conserving brain space and time, habituation also sets thresholds for knowing that something important has just happened, because it alerts us to when something has changed. When we walk down a noisy street, after a while we don't sense the constant drone of traffic passing us by, we habituate to it and keep it out of our head. But when someone blows her horn, we immediately dishabituate and are startled into alertness. Habituation and its flip side dishabituation are important adaptations for survival. It matters more to notice the sound of a branch breaking under the foot of a predator than the constant noise of wind blowing. And as noted above, brains evolve to sense what matters.

Habituation is a real problem in courtship that involves repetitive displays, and it seems that some complex courtship displays have evolved to keep the chooser from getting bored. Nightingales, for example, can produce more than a thousand songs in a single night. How do they hold a female's interest? One solution is that if you must keep singing, at least vary the melody. Even in common grackles, where each male produces only a single courtship song type in nature, females show heightened response to artificial courtship stimuli made up of a repertoire of four distinct song types.[15] Complexity is a good antidote for boredom; in the next few chapters, we will see how it has played an important role in the sexual aesthetics rooted in most of our senses.

* * *

Another cognitive process that looms large in how we perceive beauty is generalization. Just as the brain is designed to deliver appropriate mates, it is also under strong selection pressure to recognize individuals and situations that are encountered repeatedly, in order to determine if and how to respond to them. The brain is an amazing organ, but it can't know everything, and there is an important mechanism that allows us to make informed guesses when we encounter new individuals or situations—generalization.

If we have never seen an individual before, we can easily tell if it is a human or some other primate, and we often know if it is man or a woman, a boy or a girl. These discriminations are based on generalizations from what we know about our species, our sexes, and about cultural gender. Sometimes our generalizations are quite accurate, while other times they can be horrendously wrong. On average, though, generalizing is usually better than making a random guess. These generalizations can set the stage for the emergence of sexual beauty.

A fundamental skill needed for sexual reproduction is the ability to tell males from females. In many animals this is learned, and in zebra finches, learning how to identify sex leads to some interesting biases in sexual attraction. Biological parents include one male and one female. If both parents raise their offspring, this provides a good opportunity for the young to learn how to distinguish the sexes. A nestling zebra finch learns to associate the orange beak of its mother with "female" and the red beak of its father with "male."[16] When it becomes an adult, the bird then uses this information it derived from experience with only two individuals, mom and dad, to distinguish the sex of all other zebra finches it encounters in its life. But oranges and reds vary and will not exactly match the color of mom's or dad's beak. Since the brain does not have a directory that associates every shade of red and orange with every individual it might ever meet, it makes its best guess based on what it does know. If a beak is more orange, the bird is more likely a female; if it is redder, then it is more probably a male. For intermediate shades between red and orange, the finch will be more likely to make mistakes. As we will soon see, this generalization rule not only allows correct identification of the sexes but sets the stage for the evolution of more extreme differences in beak colors between the sexes.

Sometimes it's more important to be identified as who you *are not* than who you *are*. Let's consider a male zebra finch shopping for a mate. He could choose one whose beak color is the most *similar* to his mom's.

But he could also choose a finch whose beak color is the most *different* from his dad's. "Looking for mom" involves searching for a specific shade of orange, while "looking for not-dad" results in an open-ended preference for more orange. This is what the male zebra finch does.[17] The phenomenon is called *peak shift displacement*, and it can lead to the evolution of more extreme traits—in this case maybe we should call it "beak" shift displacement. Even though there might be no inherent differences among females other than the shade of orange in their beak, generalization results in the evolution of more orange female beaks to evolve because they are more reliably identified as females and better match the male's sexual aesthetics.

There are also hints of this phenomenon in our own species, as we often find more attractive those men with more masculine features, such as broader shoulders and deeper voices, and those women with more feminine features, more pendulant breasts and hourglass forms. Generalization and the open-ended preferences they generate are two psychological mechanisms that can go a long way toward explaining the evolution of elaborate sexual beauty.

* * *

Finally, we will consider a cognitive process that biases our perception of differences in magnitude—Weber's Law. In both humans and animals, we often refer to the sexual marketplace where choosers do comparative shopping for courters, and most of these comparisons are based on the magnitude of the courter's sexual traits. Hundreds of studies have shown that females often prefer males with more: brighter colors, longer tails, more complex songs, and bigger antlers.[18] But how different do these traits have to be in order to be perceived as different? More precisely, what are the rules we use in comparing stimulus magnitude? Knowing how the brain compares things can help us understand what a chooser finds more beautiful, and it might be that there is a simple law that explains a lot of this preference and the limits over which it can operate.

Human comparisons of "how much" are often based on ratios rather than absolute differences; this was first posited by one of the early psychophysicists, the German scientist Ernst Heinrich Weber, in the 1800s.[19] If we can barely detect the difference between a one-pound weight and a weight only one ounce heavier, there would need to be a much greater difference than an ounce for us to distinguish a one hundred–pound weight from a heavier one.

Weber's Law could act as a cognitive brake to slow down the evolution of exaggerated sexual beauty, at least in one species.[20] As noted in the preceding chapter, female túngara frogs prefer mating calls with more chucks compared with those with fewer chucks. If this preference is based on absolute comparisons, then the total number of chucks should not matter, only the absolute difference in the number of chucks. Thus, the probability that a female would prefer a two-chuck call to a one-chuck call should be the same as her likelihood to prefer a six-chuck call to one with five chucks. If females followed Weber's Law, however, then the preference for calls with two versus one should be stronger than the preferences for six over five chucks. When Karin Akre, myself, and a team of others tested females with pairs of calls with more versus fewer chucks, it became clear that females followed Weber's Law and used proportional rather than absolute differences in mate choice.[21] Since females can barely tell the difference between five and six chucks, there is little pressure on males to evolve the ability to produce more chucks than this, and thus the rate of evolution of additional chucks should slow down.

The frog-eating bat is also attracted to the túngara frog call. Like the female frogs, it prefers more chucks to fewer chucks, but unlike the females, the bats are looking for a meal and not a mate. We did the same choice tests with the bats that we did with the females, giving them choices between pairs of calls that varied only in the number of chucks. The bats also followed Weber's Law. Our interpretation of these results is that the way in which female túngara frogs compare the number of chucks is not an adaptation for mate choice peculiar to this species but results from a way of comparing stimulus magnitude that evolved sometime in the past before the sounds of chucks punctuated the tropical nights of much of Central America.

* * *

Up to this point, I have discussed how the brain forms an animal's sexual aesthetics, how our percepts of beauty are designed to deliver us good mates, and also how these percepts derive from other things the brain is designed to do. But it is important as well to understand not only why we see some individuals as so *sexually beautiful*, but also why we find them so *sexually desirable*. In the vernacular, we equate attraction with liking, and we often assume that liking and wanting are the same thing. But here we are wrong. "Wanting" flows from "liking," but they are not synonymous. To untangle these two concepts, we now have to probe deeply into the brain to uncover its centers of pleasure.

New Orleans is known as the "Big Easy," a nickname that contrasts its easy, laid-back lifestyle with the more hectic one in the "Big Apple," that is, New York City. What could make life easier than just pushing a button when you want to have some pleasure? Sounds like a fantasy, but it is a fantasy made real by Robert Heath, a psychiatrist at Tulane University in New Orleans in the 1950s.[22] Heath implanted electrodes in "deep areas" in the brains of patients with various maladies. When he stimulated these areas, patients who were depressed to the stage of being catatonic now smiled. Heath then put some of his patients in charge of their own pleasure, providing buttons that allowed them to electrically stimulate these hedonic areas of the brain. The results were scary. One patient stimulated himself 1,500 times during a three-hour session. With this patient and all others, however, their euphoria was short-lived; when the stimulation stopped, so did the pleasure. Similar studies with rats in that same decade revealed similar results; some rats delivered these jolts of joy one thousand times an hour, even forsaking eating to do so.

Heath had probed into the brain's pleasure centers and initiated a field of research that has grown exponentially. One of the main findings is that the neurotransmitter associated with the reward centers is dopamine. Dopamine is an opioid with properties similar to opium, heroin, morphine, Percodan, and codeine, all of which bind to receptors that stimulate dopamine production—thus this neurotransmitter's pejorative association with "sex, drugs, and rock and roll" and a suite of other pleasurable indulgences such as gambling and gluttony that are correlated with increased dopamine levels in the brain. As with many findings in science, subsequent studies revealed more detail and nuances than were first appreciated. We now know that there are two components to these rewards systems, which relate to both "liking" and "wanting." Although dopamine has many functions in the brain, none of them is delivering pleasure per se, it is not involved in "liking." As Kent Berridge and his colleagues have shown, dopamine activity modulates "wanting" by "stamping-in incentive salience."[23] I will explain what this means.

We can ask people if they like a certain food, or we can just watch how their faces respond. In the Meg Ryan and Billy Crystal flick *When Harry Met Sally*, Ryan's character fakes an orgasm while dining in the famed Katz's Deli in New York City. When her sham climax is over, a woman nearby assumes Ryan's orgasmic pleasure was related to what was on her plate and tells the waiter, "I'll have what she's having." Okay, it is not quite that easy to gauge someone's reaction to food, but pretty close. We show different facial reactions to a taste of chocolate versus sour milk, or to an outstanding wine versus one that is hardly drinkable. Rodents are little different, and Berridge used the faces that mice make when given food as an indication of how much they liked it. Specifically, he measured how much they licked their lips and whiskers when they received a sugary treat.

In a now classic study, the researchers increased dopamine levels by injecting amphetamines into one of the main sites of the dopamine reward system. These "doped-up" mice showed no more pleasure than did the normal mice in response to sugary foods, but they were willing to work harder to obtain food, to run longer on an exercise wheel for a snack. Dopamine-deprived mice, on the other hand, were not willing to work for food, had little interest in feeding, but when they were force-fed, it was clear that they found food pleasurable. This study showed that dopamine is not involved in liking but rather in wanting. This distinction also explains dopamine's widely acknowledged role in addictions of all kinds: drugs, sex, gambling, and eating.

I have been emphasizing how sense organs, areas of the brain that integrate sensory inputs, and the cognitive processes that analyze this sensory information can have important consequences for how we perceive beauty. The reward system is yet another area of the brain that can evolve to couple immediate positive reinforcement to desires that enhance Darwinian fitness. It is also a system that can be exploited by those who want to be desired. In fact, in 2015, Sprout Pharmaceuticals received approval from the US Food and Drug Administration to mass-market Flibanserin, which boosts a woman's sexual libido, her sexual "wanting," by boosting levels of dopamine and its close relative norepi-nephrine, which also enhances "wanting," and suppressing serotonin, a neurotransmitter known to inhibit libido.[24]

* * *

Now that we are familiar with some fundamentals of the sexual brain, we will delve into the details about how it perceives beauty through each of three major senses. Since humans are so visually attuned, we will start there. If I tell you that the rainforests of BCI are beautiful, you will assume I am talking about visual features such as the lush green and dappled sunlight, even though the sounds of birds, insects, and frogs and the smells of flowers and ripe fruit contribute dramatically to the appeal of this landscape. Our need to visualize individuals or things in order to know, understand, and assess them is not unique—many animals depend on vision to make important decisions like with whom to mate—and thus visual beauty is rampant throughout the animal kingdom. The next chapter will provide some insight into the oft-asked question: "What does she see in him?"

FOUR

||||||||||||||||||||||||||||||||||||||||||||||||||||||||||||||||||||||||||||||||||||

Visions of Beauty

. . . if eyes were made for seeing,
Then beauty is its own excuse for being.
—*Ralph Waldo Emerson*

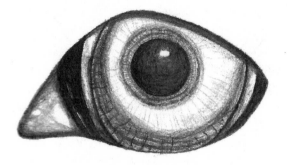

TOMMY, THE MAIN CHARACTER in the eponymously named rock opera by The Who, pleads to the minions at his summer camp to see him, feel him, touch him, heal him;[1] this naturally first appeals to their sense of vision. Although endowed with a varied sensory tool kit, we are primarily seeing animals. We are so invested in our eyes that we recruit our sense of vision in idioms that have nothing to do with vision; "seeing" is a metaphor for understanding: "Do you see what I mean?" "He can't see the forest for the trees." "She can't see straight."

One of the things we see, and often take special joy in seeing, is color. Why else was I so moved by that spectacular rainbow arching over the Irish coast? Another important attribute of our visual scene is pattern, which is especially important in our appreciation of certain types of art, such as abstract expressionism. This chapter begins by taking a look at (yet another visual idiom) how biases in our visual system for color and pattern are exploited in some human domains that are unrelated to sex. This will then set the stage for examining how these and other visual biases influence animals' sexual aesthetics and drive the evolution of sexual beauty to ends that are sometimes quite artistic.

Much of the beauty we admire is alive with pulsating colors, whether they be the blue and green chromes in van Gogh's *Irises*, the bold blacks and reds favored by Oaxacan rug weavers, or the stunning assemblage of yellows, reds, and oranges in a New England autumn. Although I am pretty sure that a female quetzal does not experience exactly the same excitement I reported in chapter 1 when I first viewed a male of her species, we both shared the experience of having our aesthetics tweaked by his collage of light greens, bright reds, and iridescent blues.

In contrast, let's consider someone with whom we have a closer evolutionary affinity than the quetzal, the New World howler monkey. The monkey's name derives from its long-distance howl, which reverberates throughout the forests of much of the New World tropics. This howl can be elicited by the calls of other howler troops; my daughter Emma, who has an uncanny ability to duet with them; and, oddly enough, claps of thunder that punctuate the rainy season in tropical rainforests. These monkeys are well adapted for howling. The howl can be heard over several kilometers, in part because it is resonated by a hollow bone near the vocal cords.[2] This bone is twenty-five times larger than in non-howling monkeys of similar size. Howling, however, is the only activity that seems to excite these monkeys. These are not energetic animals. According to Alexander von Humboldt, the famed explorer, "their eyes, voice, and gait are indicative of melancholy."[3] Although they eat fruits, I have mostly observed the mantled howler monkey in Panama grazing on leaves, and I have always wondered if their melancholy, or at least their lethargy, was a symptom of ingesting large amounts of chemical compounds that leaves contain to ward off herbivores. Although their howl is awe inspiring, there is not much stunning about their appearance. The mantled howler monkey is a monochromatic black, lacking any distinguishing hues. Yet, except for humans, it is the only other mammal in the New World in which both sexes have color vision.

The process of vision starts in the eyes, in the photoreceptors of the retina to be precise. There are two classes of photoreceptors, rods and cones. Rods allow us to see at low-light levels, while cones give us access to color and the beauty that comes along with it—without cones, van Gogh's irises would fade to a grayish tint. We share the gift of color vision not only with howler monkeys, but with most Old World primates, including all of our closest relatives, the great apes. We can see color because not all of our cones are the same; different cones are more likely to be excited by different wavelengths of light, which we perceive as different colors. Not only do we need cones for color perception, we need at least three different types. Our cone types are referred to as short-, middle-, and long-wavelength cones. These cones are most sensitive, respectively, to the colors (and wavelengths) blue (419 nanometers), green (531 nm), and red (558 nm). Most other mammals are dichromats and thus are not capable of true color vision; lacking a long-wavelength cone, they cannot perceive differences between red and green. We are trichromats, which means we have access to the wonderful world of living color to which most other mammals are blind.

Color vision is thought to be uncommon in mammals because our distant ancestors were nocturnal, as are many mammals today. Thus we assume there was little advantage for color vision, including the ability to discriminate red and green that comes along with three-cone types. For the howler monkey and other trichromatic primates, color vision in general and red-green discrimination in particular are important sensory tools. When it is lumbering through the canopy in search of fruit, the howler is surrounded by green. Color vision, it has been suggested, evolved in howler monkeys and other primates to enhance their ability to find fruits, which often have a reddish hue, against a green background.[4] In addition, for the handful of monkeys that eat leaves, it is important to recognize that not all leaves are the same. Younger leaves, which often flush out as red, have higher nutritional value than older leaves; the ability to distinguish between young and old leaves is enhanced by color vision. So when you stop at the red light and go on the green, you owe it all to your foraging primate ancestors. And when we are mesmerized by the hues of quetzals or the colors dripping from a tangled Jackson Pollock painting, we owe these joys in part to the photopigment evolution that occurred back in an ancestral monkey to carry out tasks that having nothing to do with aesthetics.

Recently, another potential advantage to color vision has been suggested as the driving factor in the evolution of primate trichromacy. This idea is fairly radical, but even Darwin spent a good deal of time studying it. In a book dedicated solely to animal behavior, *The Expression of*

Emotions in Man and Animals, he discussed the most "peculiar" of all human behaviors.[5] Guess which trait he found so peculiar—language, tool use, gluttony, or drinking milk from other animals? No, it was blushing. We all blush despite some assertions to the contrary, as Lou Reed acknowledged in "Sweet Jane," a song he penned and sang with the Velvet Underground.[6] The Misters Reed and Darwin are on the same page here, children are *not* the only ones to blush—don't let any evil mother tell you otherwise.

Darwin made use of his extensive social-scientific network to gather data about blushing. His correspondents informed him that although children did blush more than adults and women more than men, blushing was a characteristic of all ages (after infancy), of both genders, in all cultures, found in peoples inhabiting all geographical areas, and is mostly confined to areas of the neck and face. There were some exceptions to the latter. Some of Darwin's medical-doctor informants reported that blushing of some women extended to areas near the breast. Darwin noted that blushing was correlated with certain emotional states, such as shame and embarrassment, and, most critically, that we had no control over it. Blushing seems like a pretty important cue, and one might think we have visual adaptations to tap into it, much like the howler's adaptation to find fruit.

In his book *Visual Revolution*, the theoretical neuroscientist Mark Changizi argues that human color vision provides us with subtle insights into a person's emotional state, and here he means "insights" literally as well as metaphorically.[7] He utilized an analytical model that compares the tuning of our cone receptors to the color changes that take place in the skin during blushing. There was a strong correspondence between the two. The tuning of our photoreceptors, he concludes, should make us pretty good at detecting blushes. In fact, Changizi further argues that the evolved function for primate color vision is to detect slight changes in skin color brought about by changes in blood circulation, which might indicate emotional reactions to social situations as well as other physiological processes, such as physical exertion. When we met at a conference on Sensory Exploitation and Cultural Attractors in Belgium, he further explained that his company, 2AI Labs, is working on devices that will be capable of dynamically measuring changes in skin color that are below the radar of our own blush detectors. We are all pretty good at reading emotions from faces, but assessing subtle changes in skin color could allow us access to emotions that normally remain hidden from view.

Color is only one aspect of our visual scene, including our visual sexual scene. Another is pattern. We know that the visual systems of other animals are more sensitive to and react differently to certain shapes. In a classical experiment by the neuroethologist Jörg-Peter Ewert, toads would lunge at a simple horizontal line resembling a worm that passed in front of them, but they ducked their heads in fear when the same line was in a vertical position, resembling a snake about to strike.[8] Follow-up studies detailed just how the wiring of the visual system elicited these different behaviors. David Hubel and Torsten Wiesel pushed the study of visual pattern recognition to new heights, for which they were awarded the Nobel Prize in 1981. Their studies of the visual system in cats began with the demonstration that single cells in the brain respond to contours of specific orientation.[9] One particular advantage of cats' pattern-recognition system is that it makes them very sensitive to edges and thus keeps them from walking off cliffs. These and numerous other studies of visual scene analysis demonstrate that, just as the retina is not equally sensitive to all wavelengths, the visual system is not equally sensitive to all patterns.

We would assume that all animals, including us, would evolve sensitivities to patterns in nature that matter biologically, those that have some influence on our fitness. Right now you are looking at one of the most important patterns that we, as humans, have to recognize—the shapes of letters used in written language. But writing evolved relatively late in the human lineage, after our brain evolved its sensitivity to the visual scenes around us. Does that mean that the brain's sensitivity to visual patterns cannot be biased to the shapes of our letters? Not necessarily. Changizi and his colleagues made the argument that the shapes of letters should be drawn from patterns most common in visual scenes, since these patterns are the ones to which our brains should be more sensitive. When he and his colleagues examined the alphabets from many languages, they found that not all possible shapes have been recruited for letters, and that some shapes are more likely to be used than others. Those most commonly used, such as a capital or lowercase T, are the types of shapes that abound in the natural scenes around us. In fact, the average number of strokes per character in the ninety-three speech/writing systems they surveyed was three, which is pretty close to the average in natural visual scenes. It did not matter if the system had 10 letters (e.g., Mangyan, Gurhmiki, Arabic) or more than 150 letters (Dene, International Phonetic). If needed, languages would add more kinds of strokes to make new letters rather than adding more strokes per letter. The most striking result, however, is that when the researchers plotted out the occurrence of nineteen structures in human visual signs compared to the frequency of similar images in natural scenes the correspondence was almost perfect. For example, the shapes T and L are most common both in alphabets and in nature, while an asterisk (*) is least common in both signs and scenes. The visual system and the letters we view with it, Changizi argues, are tuned to one another, but it was culture that tuned letters to match the visual brain, which had already been tuned to view the natural scenes it resided in. Although we use letters to compose love missives, this fascinating work is not directly germane to sexual aesthetics, but it is a prime example of the general phenomenon of how signals evolve to exploit preexisting biases in the brain.

* * *

In this book, we are concerned with understanding sexual aesthetics in animals, including humans. In something of a parallel universe, scholars have been asking a similar question: what determines our appreciation for art, especially for paintings? The holy grail in each field would be a simple prediction, or better yet an equation, for what would be deemed beautiful. As David Rothenberg recounts in his book on evolution and art, *Survival of the Beautiful*, such an equation for art was proposed in 1933 by George Birkhoff: $M = O/C$, where M is the measure of aesthetic appreciation, O is order, and C is complexity.[10] Not surprisingly, this equation did not cause a paradigm shift in the fine arts, and it had no influence on how artists put brush to canvas, nor did it inform aesthetic preferences of those who view art. And I can guarantee you that no simple equation would work in animals either.

The reason it is difficult to predict the details of sexual beauty among animals is that the differences between the brains of species, and even between individuals of the same species, can promote numerous and idiosyncratic variations in sexual aesthetics, which in turn drive the diversity of sexual beauty. Nevertheless, there are some general themes across species and sensory modalities. The most ubiquitous is that there usually is a preference for traits of greater magnitude and complexity. There are also discernible patterns of beauty within groups of animals relying on the same sense modality. For example, the sensitivity of photoreceptors in the eyes of fishes often predict courtship colors that are preferred in mate choice, and the tuning of the inner ear of frogs predicts the pitch of the call that females find most attractive. But as with human aesthetics, there is no simple equation that explains the diversity of sexual aesthetics and resulting sexual beauty in animals. Instead we need to understand how the chooser's brain perceives sexual traits, not only those traits that exist but those that could exist; that is, we need to be able to ascertain preferences that are hidden.

When we survey the variety of sexual beauty in nature, what we see are those traits that have passed muster, those that are attractive enough to be maintained in the genome generation after generation. What we do not see is the graveyard of attempts to be beautiful, traits that mutated and made their bearers ugly rather than sexy. Studies I have reviewed up to now, and others that I will address in the rest of the book, show that there are hidden preferences for sexual traits yet to exist in a species. The evolution of sexual beauty is an ongoing experiment in every species; variation in courtship traits arise and are judged by choosers who quickly relegate most of them to the dustbin of history. But those new attempts at beauty that pass muster, that trigger one of these hidden preferences, hit the evolutionary jackpot. In the next section, we will see how this has happened with colorful traits in fishes.

* * *

If I told you that the most productive forest in the world is in California, you might assume I meant the redwoods. You would be wrong. The forest I am talking about is underwater. The kelp forests of the coastal waters of California have a majesty which competes with that of the nearby redwoods and have an ecological productivity that exceeds it. I visited this forest with a group of fish biologists some years ago. Molly Cummings was a graduate student then and today is an expert on the evolution of animal vision. Molly invited Gil Rosenthal, Ingo Schlupp, and me to visit her study site off the coast of Monterey, California. The kelp forest where Cummings worked is made up of long, leafy algae, and that day the entire forest was rocking back and forth so vigorously with the surge of the tides that few of us could keep our breakfasts down. But the stunning light-scape of this forest had us in its grasp.

Light can degrade by small particles in the environment that can have profound effects on ambient lighting. This is what happens when droplets of the water diffract light of different wavelengths to produce rainbows, and thus why rainbows are so common in the misty heather lands of Ireland. Diffraction of light in the kelp forest creates a kaleidoscope of variation in ambient light; in one open patch near the surface we were surrounded by bright blues, while only a bit deeper down the water had a distinctly red hue. When we descended farther, we were in a world of green light. We had plenty of company down there, including the animals in which we were interested—surf perches. These are denizens of the kelp forest, and different surf perch species live in different light environments within the forest. Although these fishes all eat similar prey, they have to find their quarry in different backgrounds of light. There are two general ways to detect objects against their backgrounds: one is to compare the colors of the target with the background, and the other is to compare the brightness of the target and background. The more the target contrasts with background, the easier it is to detect the target. These detection strategies are described by a burdensome jargon in visual ecology, but I'll just refer to them as color and brightness detection. Different light environments will favor different prey-detection strategies, and Cummings showed that some species have their photoreceptors tuned to maximize the target-background contrast in color, while the tuning of photoreceptors of other species maximizes the contrast in brightness.[11]

What does this have to do with sex? Enter the males. The first step in getting a mate is to be seen by her. Males evolve courtship colors to communicate with her, but like the tree falling in the forest that is never heard, a male's visual displays are for naught if they can't be seen—contrast is at a premium. How should a male maximize his signal against the visual noise in the background—enhance color contrast or brightness contrast? It depends on the detection strategy females use to find prey. In those species that use color contrast for prey detection, the males evolve courtship traits that maximize their color contrast but not their brightness contrast to females. Alternatively, in those species that use brightness contrast for foraging, the males evolve courtship traits that maximize their brightness contrast to females at the expense of their color contrast. This is an outstanding example of how selection processes that work on the visual food-detection system influence the aesthetics of the sexual brain, which, in turn, drives the evolution of sexual beauty.

As noted above, color is only one attribute of visual objects; pattern is another. Some patterns in nature are static, such as the stripes on some species of surf perch and the patterns of letters that were discussed above. But others are dynamic. Let's now take a look at how the dynamic nature of vision can influence the dynamic patterns of courtship that are considered beautiful. In this case, the female is the courter, and the male the chooser.

One function of vision is to know where you are going, and this is more easily achieved if your perception of the world passing you by is smooth rather than jumpy. A critical parameter of the visual system for perceiving movement is the flicker fusion rate. This is the rate at which a light stimulus appears to be constant. Movies and television present a series of still images in rapid succession to achieve the perception of constant movement. The human flicker fusion threshold is about 16 cycles per second (or hertz, Hz). Movies are usually recorded at 24 Hz, and TV usually at either 25 or 30 Hz, substantially above our flicker fusion threshold. If the images are presented below the flicker fusion rate, as are some older films and animations, movement is perceived as jumpy instead of smooth.

Back to sex. Although all insects use their eyes to see where they are going, some also use their eyes to choose mates. The male fritillary butterfly is a case in point, and one in which females court while males choose. Females remain stationary and flap their wings to attract males. The faster they flap, the more attractive they are to the males. The typical flapping rate of a female fritillary is about 10 Hz. In the middle of the twentieth century, the German biologist D.E.B. Magnus created a flapping machine that could mimic the female's flapping at any speed.[12] He showed that faster is better, as males preferred a flapping rate of 10 Hz to a slightly slower one of 8 Hz. If 10 Hz is sexier than 8 Hz, would 12 Hz be even sexier, and if so what about 20 Hz?

A related question is why don't females flap faster? There are two possible explanations. One is that the preference of the males does not extend above 10 Hz; perhaps higher speeds are merely as attractive, or even less attractive, than 10 Hz. The other possibility is that there are mechanical limits to how fast a female can flap. To drill deeper into this question, Magnus shifted his flapping machine into high gear and let it fly. He showed that males always preferred the higher flapping rates, even when they were supernormal; that is, faster than the flapping rate that typically occurs in nature. At least that was true up to a limit, and that limit was very fast: 140 Hz. Humans and the butterflies alike can perceive the individual flaps at 10 Hz since it is below the flicker fusion rate for both of us. If Magnus ran his supernormal flapping tests with humans and asked us to choose the faster flapper, we would reach a threshold at about 16 Hz; 18 Hz and 25 Hz would appear similar to us. As they are both above our flicker fusion rate, the flaps would be perceived as one continuous motion. Insects generally have much higher flicker fusion rates than we do because they travel through their environment at much faster speeds and need to perceive the details of the optic flow of the environment whizzing by. Can you guess the flicker fusion rate of the male fritillary butterfly? You got it, 140 Hz!

Magnus's exploration of the sexual aesthetics for flapping shows that male fritillary butterflies have a fairly open-ended attraction for flapping rate: the faster the courtship display the better, as long as the choosing male can see it. The 140 Hz speed limit on flapping preference is not dictated by sexual selection; instead it is a result of natural selection favoring a visual detection system that allows rapid travel through the environment. Because the flicker fusion rate is so high, certain sexual preferences of choosers are now in place, even if courters can't match them. But if there were a major mutation that redesigned the aeronautics of butterfly flight to allow quicker flapping, then the advantage to the fastest flappers would be immediate. There would be no need for the choosers to evolve a new preference for a new, faster speed of flapping; the preference would already be there.

We can think of a human analogy with cars and speed limits, in which the process is reversed. Most automobiles today can exceed most if not all speed limits on the roads we drive. Current engineering technology would allow the design of typical autos to go even faster—racing cars prove the point. But existing speed limits diminish the usefulness of high speed for everyday use. We would not expect a big boost in speed until we see a big boost in speed limits. In the butterfly example, however, the constraint is reversed. The speed limit for flapping is 140 Hz, an order of magnitude above the speed achievable by the butterfly. It would be as if the speed limits on our highways were much faster than any of our cars could drive, which we assume would lead to engineering innovations in the automobile industry. So we see that the butterfly system is one poised for engineering innovation to increase flapping speed. The fact that it has not happened suggests that there are other limits, probably some very basic biomechanical ones, that are constraining these innovations.

* * *

We have been considering how some basic visual processes, such as detection of color, brightness, pattern, and movement, can drive the evolution of beauty. Now we will move on to some higher processing biases that are equally important to the visual-sexual aesthetics of animals and humans alike.

In chapter 3, we examined two cognitive processes that can influence sexual aesthetics, Weber's Law and peak shift displacement. Weber's Law dictates that decisions about quantity are based on comparisons of proportional rather than absolute differences. So as traits get larger, such as the peacock's tail, the difference between them needs to be greater to be perceived. Thus Weber's Law could put a "cognitive brake" on the evolution of extravagant beauty: the bigger the trait, the less likely a slightly bigger one will be favored by females.

The other cognitive driver of beauty discussed in chapter 3 was peak shift displacement. We reviewed how zebra finches use their parents' beak color to identify sex, red for males and orange for females, and how peak shift displacement causes preferences for those traits that are most different from the same sex—males prefer beak colors that are most different from dad's and thus more likely to be a female. Peak shift displacement can give rise to open-ended preferences and favor supernormal traits, similar to the one we just saw in the fritillary butterfly. Just as the male butterflies favor flapping rates out of the reach of females, male zebra finches might prefer shades of orange beaks that females could never evolve. Although peak shift occurs through learning and can lead to preferences for supernormal stimuli, it does not follow that all supernormal preferences arise from peak shift. The butterflies' open-ended preference, for example, does not involve learning but has more to do with the rate of stimulation of visual neurons. But regardless of what generates it, preferences for supernormal stimuli are important drivers in the evolution of sexual traits that are bigger, brighter, and faster. We will turn to one compelling example now.

I once visited Kenya with my colleague Merlin Tuttle to study the heart-nosed bat, *Cardioderma cor*. Like *Trachops*, this bat eats frogs, but we found that unlike *Trachops*, its auditory system has not been retooled by evolution to locate frog calls. During that trip we had two close brushes with death—bandits and elephants, which Merlin recounts in *The Secret Lives of Bats: My Adventures with the World's Most Misunderstood Mammals*.[13] A more rewarding encounter occurred in the highlands, where I spotted an animal that must have a fatal close encounter every day of his life. A bird flew across my field of vision only slightly above the tall grass. Until I could focus, it was a confusing sight. The bird was small and mostly black. Its wing span was only twelve centimeters or so, and it was closely followed by something much larger, maybe a half meter long, that stayed right on the bird's tail. When I finally focused, I could see that the object behind the bird was its tail, one that was several times longer than the rest of its body. It was a widowbird.

It didn't take long for me to realize how this bird got its name. I remember listening to a baseball game once when the pitcher threw a fastball close to the batter's head. The announcer called the pitch a "widow maker"; the implication was obvious. More tragically, during the Troubles (1968–98) in Northern Ireland, the weapon of choice of the Provisional IRA was the Armalite 18, an efficient killing machine originally designed for the US Army, which got into the hands of gun-runners—it went by the nickname "widow maker." I was sure that the long tail attached to the little bird was also a widow maker; given the bird's burdened flight, it was clear its days were numbered. This ornament is not a "widow*er* maker," as only males have the excessively long tails. All of this must have been clear to the Dutch physician and naturalist Pieter Boddaert, who bestowed the moniker *widowbird* on this species in 1783.

The widowbird's tail is truly remarkable, and it has evolved in response to a sexual aesthetic that just wanted more, more, more. This is exactly what Malte Andersson, who wrote the classic *Sexual Selection* in 1994,[14] had earlier demonstrated in one of the definitive experiments in sexual selection.[15] He measured the mating success of male widowbirds by counting the numbers of nests on their territories on the Kinongop Plateau, about one hundred kilometers north of Nairobi, where it is one of the most common birds. He then sought to discover if the length of each male's tail would lead to differential mating success. To get at this, he did a "cut-and-glue" experiment. In one group he cut the males' tails to make them shorter, he then glued the cut portions of those tails to males in another group to give them tails that were extra-long, or supernormal. The third group was a control: he cut off a portion of the male's tail but then glued it back on. The next month he again measured the mating success of each male and compared it to the same male's mating success prior to the experiment. Males with supernormal tails had increased mating success, control males showed no change, and males with shortened tails had decreased mating success. Andersson showed that an open-ended preference for supernormal tails drives the evolution of tail length, and, although not proven, it seems a good possibility that death slows down the evolution of tail length as longer tails make more widows.

Not all preferences, of course, are open ended. Some are for specific patterns.

* * *

As with many occupations, science has its get-togethers. They can be small or immense; for example, the Winter Animal Behavior Conference that I attend is limited to thirty participants, whereas the meetings of the Society for Neuroscience usually hemorrhage at thirty thousand. Regardless of size, these conferences are good places to find out about groundbreaking research before it hits the presses. I was at a reception at such a conference in Kyoto, Japan, in 1991 when two well-known behavioral ecologists, Randy Thornhill and Anders Møller, gave me the scoop about their radical idea for the evolution of beauty—fluctuating asymmetry. This is what they suggested. We, and most other animals, are bilaterally symmetric. If we draw a line down the middle of our body, the left and right sides are mostly the same, such as the length of arms, legs, and fingers. There are exceptions, of course. Male fiddler crabs have one extremely large claw and one smaller one; for males of a given species, the same claw, right or left, is the larger one. (It is more difficult to think of animals whose entire body plans are asymmetric; here's a hint about one of them: think of the eponymous cleansing device you use in the shower and sink.) Most other exceptions to symmetry, however, are small deviations called *fluctuating asymmetries* (FAs), which are equally likely to occur on the left or the right. When animals are stressed during development, they have larger FAs. Thornhill and Møller's idea was that animals with superior genes for survival should be buffered against developmental stressors, have lower FAs than less well-endowed individuals who develop under the same stress, and, they predicted, females should have a preference for more symmetric males and the good genes they carry.[16]

The idea of sexual selection and FA sparked a profusion of studies asking if symmetry was a key ingredient in sexual attractiveness, and if so, why. The answer to the first question was yes; it appears to be a criterion in the sexual aesthetics of a number of species, especially birds and humans. For example, Møller conducted a cut-and-glue study similar to the one with widowbirds, but he manipulated the symmetry of tail feathers in swallows.[17] As he predicted, females preferred more symmetrical males. Thornhill and his colleagues showed that preferences for symmetry influence human perceptions of sexual beauty as well. Legions of researchers have shown that we are often more attracted to more symmetrical faces, and Thornhill even showed that women with more symmetrical partners have more orgasms during intercourse than women with partners who are a bit out of whack. Of course, in the field of beauty there is variation in the eyes of the beholders; other studies have shown that humans sometimes judge symmetrical faces as less attractive than asymmetrical ones. There are exceptions in other animals as well; my own studies of cricket frogs showed that symmetry played little role in influencing attractiveness.[18] If a preference for symmetry in mates is nearly universal across taxa (and this is a big *if*), are good genes the only explanation for such a preference; that is, did preference for symmetry in courters evolve because of the genetic advantages accrued by choosers for their offspring? Or does the basis for this preference lie in some other domain?

In chapter 2, I showed how the preference for chuck number in female túngara frogs followed Weber's Law. There were two hypotheses that could explain why females have this particular pattern of preference. The first was that this preference evolved in túngara frogs because relative chuck number is a good indicator of relative male quality. The second hypothesis was that this preference results from a perceptual or cognitive bias—this is just how brains work; there is no need to invoke sexual selection to explain this preference. The fact that frog-eating bats also followed Weber's Law in their preference for chuck number supported the cognitive bias hypothesis. A similar discussion has arisen about symmetry preferences: did they arise under selection favored by the advantages of getting superior males, or were they the result of a more general perceptual or cognitive bias?

One argument for a cognitive bias underlying symmetry preference is that it occurs in a variety of animals in domains having nothing to do with sex. We have preferences for symmetry in certain kinds of art, architecture, interior designs, flowers, and pets, as well as faces. Bees excel at learning symmetrical patterns over asymmetrical ones, and they prefer to pollinate flowers with more symmetrical petals. In fact, even chickens prefer symmetrical human faces to asymmetrical ones.[19] The correlation between preferences of chickens and humans for the same human faces that varied in symmetry was 98 percent! But if there is a cognitive bias, where does it come from?

This issue of a cognitive bias was first addressed by studying symmetry not in birds or humans but in computer brains. The "brains" here are artificial neural networks (ANNs). The networks consist of computational units that act like neurons and are connected into networks that mimic nervous systems. And like nervous systems, stimuli can be input into the system, and "neural" responses come out the other end. These models have a wide range of applications, including pattern recognition, stock market predictions, and traffic management. My colleague Steve Phelps and I have used them to model brain evolution.[20] ANNs were critical in alerting FA researchers to the possibility of symmetry preferences emerging from cognitive biases.

Two biologists, Antony Arak and Magnus Enquist, trained ANNs to recognize objects that were asymmetrical; to do this, the dynamics of individual neurons in the network were tweaked until the output was greatest in response to the training objects.[21] Once the ANNs were trained, they were then presented with novel objects that were symmetrical as well as asymmetrical. The networks actually showed a greater response to the symmetrical objects, even though they were trained to prefer asymmetrical ones. This means that, at least in ANNs, a preference for a novel symmetrical trait can emerge as a result of learning to prefer other kinds of asymmetrical ones, supporting the idea that symmetry preference can result from a cognitive bias.

John Swaddle, who coauthored *Asymmetry, Developmental Stability and Evolution* with Møller,[22] visited my department in Austin to lecture about a very different research topic—how noise pollution affects birds. But he still pays close attention to the world of symmetry. Over lunch, Swaddle argued convincingly that symmetry preference is a by-product of how we perceive shapes, and his work on starlings showed the same results as the ANNs studies, that symmetry preference can emerge as an offshoot of general learning phenomena. But why is this the case?

One example of such a bias comes from a theory about prototype formation. The idea is that the average of a bunch of randomly asymmetrical patterns is a symmetrical one. Most of us have one leg that is slightly longer than the other, but since the longer leg is as likely to be the right one as the left one, the average difference in leg length is pretty close to zero, and when we imagine an unknown person, we imagine him or her with legs of the same length. Thus, the mental image or the "prototype" that emerges from being trained with asymmetrical objects is the average of these objects, which will be symmetrical. After learning, a symmetrical object then best matches the prototype. This can explain the results for the ANNs, starlings, and chickens. It seems that preferences for symmetry in sexual traits might have nothing to do with good genes of the courter but more with how the brains of the choosers work.

It is possible that symmetry preferences could deliver genetic benefits to choosers even if that is not why those preferences evolved. As Thornhill and Møller suggested, symmetric individuals might be genetically superior when it comes to overall health and vigor. Symmetry preferences then could, in theory, deliver genetic benefits to choosers by endowing them with healthier offspring. Whether or not symmetry preferences are beneficial to the chooser in mate choice, however, these preferences still would drive the evolution of symmetrical courtship traits because it is part of the chooser's sexual aesthetic, even if it is an incidental one. Once again, we see that part of the aesthetics of the sexual brain might originally have had nothing to do with sex.

* * *

We often think that people are born with their looks. The beautiful have won a genetic lottery, and there is little the losers can do to improve their fate. Cameron Russell is a high-fashion model whose good looks have landed her on the covers of *Vogue* and *Elle* and opened for her the runways at Victoria's Secret and Chanel. But Russell has declared that looks are not everything, and she has become known as the "renegade model" for her criticisms of how the media contributes to problems of self-image for many young women. In a widely viewed TED Talk, she stuns the audience with before-and-after photos in which her image transforms from an innocent-looking young teenager to a sexually alluring vixen. "And I hope what you're seeing is that these pictures are not pictures of me. They are constructions, and they are constructions by a group of professionals, by hairstylists and makeup artists and photographers and stylists and all of their assistants . . . and they build this. That's not me."[23] Perhaps the lady doth protest too much methinks, because Russell, as she acknowledges, did win a genetic lottery. But she has been further engineered to be even more attractive, sexual, and stunning and, consequently, wealthy beyond most of our dreams. But in that sense, she is not an exception in the community of sexual animals, many of whom are able to improve on their genetic-given beauty.

Richard Dawkins wrote one of the most important books in biology in the past century, *The Selfish Gene*, where he offers us a "gene-centric" view of evolution, a world in which genes are immortal replicators and individuals merely ephemeral vehicles to transport them across generations.[24] He followed up that book with another important contribution, *The Extended Phenotype*.[25] The main idea of that work is that genes contribute to our physical makeup, our phenotype, but our phenotype extends beyond our body. It includes manipulations of our bodies as well as tokens and resources that we accrue.

No force in nature causes individuals to extend their phenotypes as does the drive to increase one's sexual beauty. Our own species is the best example of this. We all come with portfolios that contribute to our sexual beauty. The man with a healthy head of hair and well-chiseled physique looks even sexier when he hops into a Lamborghini or shows off his herd of cattle. For some men, when an attractive woman dons a pair of reading glasses, her attractiveness goes up, because they assume there is superior intelligence to complement her appealing physique. By acquiring accessories that enhance our sexual attractiveness, which in humans is often judged in monetary currency, we advertise what we have to offer a potential partner. These accessories become who we are. Animals are no different.

One way to enhance our beauty is to decorate our surroundings. "Come up to see my etchings" is a come-on that might be as old as cave paintings. Art seems to be an indicator trait for humans, as it reveals that an individual has such a surfeit of resources for fundamentals that she or he can splurge on excesses. The more art I have, the more expensive the art I have, the more you can safely assume that I am flush with riches . . . want some? The cost of the art rather than the art itself is the meaning of the message.

Animals decorate too, and for the same reason. The animal artists extraordinaire are the bowerbirds. Although the codiscoverer of the theory of natural selection, Alfred Wallace, provided some of the earliest scientific descriptions of bowerbirds,[26] it was the Pulitzer Prize–winning evolutionary biologist and geographer Jared Diamond, better known to the public for his popular books such *Guns, Germs, and Steel*,[27] who brought the "artwork" of these birds to the attention of modern science. He reports his initial encounter with bowers in *The Third Chimpanzee*: "I had set out that morning from a New Guinea village, with its circular huts, neat rows of flowers, people wearing decorative beads, and little bows and arrows carried by children in imitation of their fathers' larger ones. Suddenly, in the jungle, I came across a beautifully woven circular hut eight feet in diameter and four feet high, with a doorway large enough for a child to enter and sit inside. In front of the hut was a lawn of green moss, clean of debris except for hundreds of natural objects that had obviously placed there intentionally as decorations."[28]

This was no kid's playhouse; it was a male bowerbird's boudoir. There are twenty species of bowerbirds, and in all of these the males construct bowers and decorate them lavishly with flowers, stones, shells, and man-made objects; some even use crushed berries to paint their bowers. The only function of bowers is to create an attractive environ in which the males display to females. It is not a nest and offers no shelter from the storm. It is a most interesting example of courters extending their phenotypes in the service of sex.

Diamond had come upon the bower of the Vogelkop Bowerbird. This species decorates its bowers with various colored fruits, flowers, and butterfly wings. He was able to glimpse the bowerbird's decorating aesthetics in several simple experiments. First, he moved around the decorations in a male's bower and showed that the males always moved them back to their original position. Diamond then found that if he placed poker chips near the bowers, the males would often take them and use them as decorations, but they were picky about which colors they used. In general, they disliked the white chips and preferred the blue ones, although different males had different preferences for colors. Sometimes when Diamond added poker chips to a male's bower, a neighboring male would come and steal them for his own bower. The only reason that males decorate their bowers is to attract females; females prefer males with more decorations, and females of different species often prefer different colors. It is no wonder that Diamond suggested, "These are birds that can build a hut that looks like a doll's house; they can arrange flowers, leaves, and mushrooms in such an artistic manner you'd be forgiven for thinking that Matisse was about to set up his easel."

Only humans create more elaborately decorated structures than the bowerbird's bower. Bowerbirds have a large brain, and among species of bowerbirds, their brain size correlates with bower complexity, which, as the lead-researcher neuroscientist Laney Day pointed out, "range[s] in complexity from simple arenas decorated with leaves to complex twig or grass structures decorated with myriad colored objects." Some researchers have suggested that the details of a male's decorations—how rare they are, for instance—show off his brain power to females. Others, such as Joah Madden and Kate Tanner, suggest that males choose decorations to exploit sensory biases in females.[29] The colors of bower decorations influence a male's attractiveness to choosing females and thus his mating success. The researchers tested the notion that these mating preferences coincide with foraging preferences, somewhat analogous to the example of surf perch discussed earlier in this chapter. For the two species they tested, the more that females preferred grapes as food, the more likely males were to include grapes in their decorations. As Madden himself pointed out to me, not everyone agrees with their findings—the diversity of bowerbirds is second only to the diversity of scientific opinions about why they do what they do. Understanding bowerbirds is still a work in progress. A remarkable example of some recent progress is how some male bowerbirds use perceptual illusions in a way that would make Walt Disney proud.

As just noted, Diamond showed that when he moved around decorations of the Vogelkop Bowerbird, the males returned them to their original position. We would not hang a painting just anywhere on a wall, but why are these birds so particular about where they put their decorations? As with alphabets and symmetry, preferences for patterns are tied up with pattern perception. The decorations so carefully placed around a bower differ both in size and in the distance from the male's display area in his bower court. The size of the image that a decoration projects onto the female's retina is dependent on its size and its distance from the female. Like most of us, she can calibrate size and distance. But as we will soon see, the males can manipulate her calibration to make themselves look better. Yes, this seems vague now, but I will explain.

Objects that are farther away look smaller because they subtend a smaller angle on our retina. Our visual system "knows" this happens, and we can make a good estimate of size independent of distance. But imagine if our brains could not compensate for this basic fact. You might think the coffee cup on your desk is really larger than the looming skyscraper in the background. You might not be frightened by that grizzly bear that appears to be a speck on the horizon; you could just squash her underfoot if she gets too close. But, of course, if she gets too close, it would be too late because in reality she is much larger than a speck. We know that perceived size changes with distance, and we don't get fooled. Or at least we don't get fooled all of the time.

Artists have hijacked our joint perception of distance and size to their advantage to manipulate what we think we see. Hobbits and dwarves in a movie seem to be standing with other, larger characters but are actually some distance away to make themselves appear smaller. The Cinderella Castle at Disney's Magic Kingdom offers an example relevant to the bowerbirds. If the windows in a building are all the same true size, the images of the windows on higher floors look smaller, since they are farther away. Our brains compensate for the image-size/distance effect to give us a realistic estimate of the height of the building. But Uncle Walt has deceived us. The windows on higher floors are smaller than those below; thus they appear to be farther away than they really are, and the Castle becomes taller in our brains. This is called forced perspective. Pretty clever on Disney's part, but remember that bowerbirds have pretty large brains themselves.

John Endler is one of the most creative evolutionary biologists around. He made his career with detailed studies of the evolution of color in guppies in Trinidad. More recently, he and his colleagues discovered how the Great Bowerbird has picked up on this trick of forced perspective.[30] As with the Vogelkop, a male Great Bowerbird can be very particular about how he arranges his decorations, such as shells and bones. Males construct bowers with a long avenue. Females enter the avenue, and from this vantage point they view the male displaying from his bower court. The male arranges the objects in a pattern such that they increase in size with distance from the avenue entrance and thus are larger as they get closer to the bower. This pattern creates a forced perspective opposite that of the Cinderella Castle, making the bower seem smaller than it actually is. We can't look into the avenue through the bowerbirds' eyes and brains, but Endler and his colleagues guessed that this particular arrangement of ornaments gives the females an exaggerated perspective of the size of the male displaying on what is falsely perceived as a small bower. So no matter how large the male really is, he will appear even larger by constructing this perceptual illusion. As with the Vogelkop Bowerbird, the Great Bowerbird also places its objects back in the same places after they have been moved around by intrusive researchers.

Bowerbirds are not the only ones who decorate for sex. Some African cichlid fishes build volcano-shaped bowers in the sand that can be up to three meters in diameter, and like the bowers of bowerbirds, aspects of the bowers contribute to the male's sexual beauty. Other cichlids decorate their territories with snail shells, but unlike bowers, the shells have a more utilitarian function, as this is where females lay their eggs. The more shells on the male's territory, the more females he mates. In a bizarre case, another bird, the male Wheatear, carries stones to his nest cavities prior to when his mate lays her eggs; the stones don't attract her, since she is already there. In an average week, a forty-gram bird will carry one to two kilograms of stones to his nest—as much as fifty times his weight. The stones serve no immediate purpose, but researchers have suggested that much like guys pumping iron in the gym, these males are showing off their strength to their females. In one last example, many fiddler crabs erect pillars to add to their displays in which they wave their large claw back and forth. The vertical structure of the pillars is especially detectable given the layout of detectors in the crabs' eyes. Although in many of these cases we do not know how these extended sexual phenotypes interact with the sexual brain, my guess is that they are importantly influenced by perceptual and cognitive biases that contribute to the chooser's sexual aesthetics.

* * *

A relatively new science addresses human visual appreciation for beauty from the mechanistic perspective, neuroaesthetics. Humans have a well-developed visual aesthetic sense that we apply to numerous domains such as the fine arts, natural scenes, and, of course, sexual beauty. How does our visual perception of traits interact with our sexual brain to elicit the percept of beauty? Visual neuroaesthetics asks the brain why it likes what it sees. As the cognitive neuroscientist Anjan Chatterjee points out, visual processing can be divided into three categories: early, intermediate, and late. Early vision extracts simple elements from the visual environment, such as color and brightness, much as was noted about surf perch earlier in this chapter. Intermediate processing segregates those elements into coherent regions, and later processing determines which of these coherent regions get our attention.[31] Forced perspective in bowerbirds arises in this late stage of processing.

Our visual aesthetics can be influenced by each these processing categories, and biases in them can arise from cultural influences or be hard-wired in the brain. The preference for symmetric faces, Chatter-

jee argues, probably does not involve cultural experience, as this preference is found across cultures. In addition, the behavior of infants, which
in theory still might conform to some cultural expectations, also suggests a hard-wired preference for symmetry: within a week of being
born, infants have a preference for looking at more symmetrical faces,
and by six months of age they actively engage more attractive faces. As
we saw in symmetry preferences in other animals, there seem to be some
basic properties of the visual system that bias the sexual brain toward
preferring symmetry. The preferences of swallows for tail symmetry,
fishes for stripe symmetry, and our own preferences for symmetry in
art and faces might all derive from the same basics of how vision works.

There are numerous examples of culturally based percepts of beauty.
Darwin, for example, concluded that variation in human skin color
arose from culturally derived preferences for mates of a particular skin
color. Similarly, preferences for hair color, hair style, overall body form,
and waist-to-hip ratio are all thought to be molded by our local culture.
The nature-nurture debate, the importance of the roles of genes versus
experience, is no longer an interesting one in biology. Most traits seem
to be influenced both by genes, be it sequence differences in DNA or
regulation in gene expression, and by the world around them, both inside and outside of the body. Traits do not differ in being either "nature"
or "nurture" but by the degree to which nature and nurture interact. Regardless of where one stands on the nature-nurture spectrum—and this
debate still rages on in the social sciences—this really has little effect on
how beauty evolves to tweak our sexual brain. Whether preference for
a courtship color derives from the opsin sequence that determines color
sensitivity of photoreceptors, as in surf perch, or if it is due to learning
of their parents' beak colors, as in zebra finches, these preferences drive
the evolution of courtship color.

Our percepts of beauty are strongly influenced by our sensory systems, but they do not reside there. As discussed in chapter 1, the sexual
brain involves all the neural systems that access information about sexual beauty in the world around it, analyzes that information, and then
makes decisions, such as what is beautiful. One area in which studies
of human aesthetics surpass those of animal aesthetics is in the use of
neural imaging techniques to determine how various visual stimuli, be

they abstract art or sexual images, boost the dopamine reward system, the part of the brain that modulates "liking and wanting," which was introduced in the preceding chapter.

Numerous studies of humans have shown that viewing attractive images, be they faces or whole bodies, stimulates various areas of the brain that are associated with the reward system. But not only do we feel pleasure when we view sexually attractive images, we feel desire. The reward system is where pleasure becomes linked to desire. Not only do we like it, we want it. This is the same system that is hijacked by some drugs, foods, and gambling and that can turn some basic hedonic pleasures into disabling addictions, a topic I will further dissect in chapter 8 when we visit pornotopia. Neuroaesthetics holds great promise in being able to unravel our sexual aesthetics by understanding not only why sexual traits are attractive but why we desire them so much. It is the sexual desire that is at the base of our percepts of sexual beauty.

Neuroaesthetics studies usually gather information about subjects' response to visual images rather than to attractive sounds or smells. So much of what we perceive when it comes to sex comes through our eyes. But we also listen, touch, and smell to assess sexual beauty, and many other animals are much more heavily invested in these other sensory modalities. Next, we turn our eyes and ears to animals that care more about the sounds of their partners than their looks.

FIVE

||||||||||||||||||||||||||||||||||||||||||||||||||||||||||||||||||||||||||||||||||||||||||

The Sounds of Sex

*The song functions as affective rather than symbolic symbols, and the
variety is generated not to diversify meaning, but rather to maintain
the interest of anyone that is listening.*   —Peter Marler

HELEN KELLER REPORTEDLY OBSERVED that blindness separates people
from things, but deafness separates people from people.[1] Of course,
people in the deaf community are hardly isolated from one another,
and they forcefully reject Keller's assertion. But the different senses access the world differently and their perceptions can have quite different
textures. Sight offers one and sound another. The sounds of sex are not
restricted to the gasps and grunts of the human bedroom. Sounds constitute a major portion of courtship in both animals and humans. The
songs of birds, frogs, and crickets, the roars of red deer, the drumming
of fishes, and even a substantial portion of human music is all wrapped
up in sex.

As we discovered in chapter 2, nearly all of the six thousand species
of frogs known to inhabit our planet have mating calls that are specific
to their species. When researchers have asked, they find that females

are nearly always attracted to the calls of their own species over others, and when researchers drill deeper, they find that female frogs' brains are wired to make their males sound more attractive than interlopers from other species. The onset of spring in the temperate zone or the rainy season in the tropics often erupts with thousands of male frogs and toads singing the call of their species in an attempt to seduce females. This is what we heard both during the day and after sundown in the cloud forests in the mountains of western Panama during one visit in 1990. In the area around Fortuna there is a diurnal frog, the harlequin frog *Atelopus varius*, which has a bright green patterning on a black background. It is conspicuous to both our eyes and ears as it makes short, high-pitched whistles from atop rocks in the splash zone of the fast-flowing, cold streams that travel from the mountaintops. It is not common for frogs to call for mates during the day, but neither is it rare. These frogs are unusual, however, in that they lack ears, or at least external ears, the eardrums that are typically on the outside of frogs' heads. Harlequin frogs also lack the middle ear bones that connect the external ear to the inner ear. Given the auditory challenges of these frogs, my friends Walt Wilczynski and Stan Rand and I wondered if they could even hear, and if so, could they locate the source of a call with such a deficient auditory system?

When we visited Fortuna, harlequin frogs were so common we had to be careful not to step on them as we hiked up the streams. We conducted experiments on their localization behavior; we played them calls of intruder males from a speaker in the stream hoping to spark a fight and see how well the resident frog could locate the source of the intruder's call. To make a painfully long story mercifully short, the frogs did appear able to find the source location of the calls, but our experiments failed to give us any insights into how they did it. Our research failed quite spectacularly, but we received some succor during the night, when we were serenaded by the mating calls of hundreds of frogs of a dozen or so species; we knew that these other species of frogs could hear and that they did it the old-fashioned way, using all the parts of a normal frog's ear. We resolved that we would have to rethink our research approach to these earless frogs and come back to Fortuna some other time to delve into the mechanics of the harlequins' hearing.

That day never came and never will. Stan Rand passed away in 2005. By then, so had nearly all of the frogs on these mountains in western Panama, meeting their demise at the hands the deadly chytrid fungus that has caused the extinction of legions of frogs throughout the world. When I was told that researchers could no longer find *any* frogs in Fortuna, I was incredulous. Two friends, Tony Alexander and Steve Phelps, and I went to Fortuna to see and hear for ourselves. Tromping through the forest for days and nights in perfect conditions for frogs, light rain at night, misty clouds during the day, we searched, training our ears on the deafening silence emanating all around us. In total, we heard one single peep from one single frog. When we tracked down the call and saw him, we had the feeling we were looking at the loneliest animal in the world. Unfortunately, this is the kind of catastrophe that Rachel Carson forebode in *Silent Spring*, the book that helped launch the environmental movement.[2]

Since that time, chytrid has spread from the mountains of western Panama, across the Panama Canal, and is now heading for South America. Recently, my graduate student Sofia Rodriguez and I found that even túngara frogs in the far reaches of the Darién Gap, an undisturbed primary rainforest far from any roads that seem to facilitate chytrid's invasion, have now become infected.[3] Túngara frogs and some other lowland frogs seem to have some resistance to the fungus; although it is debilitating to individuals, there seems to be no population extinctions yet attributable to the fungus in túngara frogs. This is probably because these frogs reside in the lowlands, where high temperatures are not suitable for the fungus. Another part of the explanation might be that males unintentionally advertise their chytrid status, and females are able to discern this information. Sofia showed that the calls of individual infected males differed from those who were not infected.[4] She then tested female preferences for pairs of calls of males when they were in each of these two conditions. Females preferred the calls of males when they were healthy over the calls of the same males when they were infected with chytrid. Although I rejoice at the continued survival of my favorite species of frog despite the onslaught of chytrid, this provides little solace given the incredible loss of biological diversity that has resulted from this single microorganism.

Please excuse the slight diversion from the topic at hand, but this book is about real animals, and many of them have real problems surviving, not just in finding sex. Now we will drill deeper into the sexual aesthetics responsible for all of the lovely sounds made by these animals.

* * *

Hearing is a remarkable feat, as is vision, but they are quite different. When we speak, we vibrate the two vocal folds in our larynx, each less than a couple of centimeters in length. The vibration results in changes in the air pressure around the larynx. These changes in pressure are subsequently modulated by the resonant frequencies of our throat, and as the pressure fluctuations exit the body, they are further shaped by our tongue and lips. When a word escapes our mouth, it changes the air pressure around us as molecules become more tightly and then more loosely packed in a pattern that eventually imparts meaning to the sound. Those changes in sound pressure first initiated by our vocal folds eventually reach the heads of our targets, the persons whose behavior we want to manipulate. When they hear us, their ears might not burn, but their eardrums vibrate in response to these pressure changes. The vibrating eardrum moves the chain of middle ear bones, one end of which is anchored to the inside of the eardrum and the other to the inner ear, and these shaking bones cause the fluid in the inner ear to slosh around. When the fluid sloshes, the hair cells in the inner ear, the auditory neurons, fire and these neural responses all reach the auditory brain, where they are processed in great detail. If these are sounds of sex, the auditory system feeds these neural responses into the sexual brain. I have left out some detail here, and there is wonderful variation on the theme of hearing among animals. But you should get the general picture, or at least hear the sound track.

Let's reflect a bit on how marvelous hearing is. Imagine being blindfolded and placing your hands on the surface of a still pond as someone launches a rock into it. You would be able to tell something happened as you felt the water's surface vibrate. Maybe, if your senses were quite keen, you could guess the size of the rock, at least whether it was a pebble or a boulder, but never its color, its temperature, or who threw it. Those surface vibrations tell you very little. But if I flap these two little folds of tissues inside my larynx, I cause fluctuations in air pressure waves that can inform you about any number of things, and besides garnering all of that "intended" information I am sending, you could also make a good guess of my gender, size, and age. Also, with my flapping folds I could elicit various emotions from you, such as amusement, anger, or fear, and I could even make you approach me, avoid me, or attack me . . . or someone else.

In this chapter we will think about how by merely making sounds courters can inform choosers of not only who they are, such as what species, but what they are like, if they are young or old, fit or infirm. We will also consider how these sounds have been designed to reach deep into the chooser's brain to best influence the chooser's attentional and motivational states, her hormonal milieu, her reward systems, and eventually her choice of a mate.

* * *

We have already considered how important it is that choosers mate with courters of their own species, conspecifics, since matings with heterospecifics are usually a waste of energy. We expect, and indeed we find, that this need has resulted in courters evolving traits that identify who they are and choosers evolving perceptual biases that make conspecific courters more attractive. Thus the brain, in this case the auditory brain, should evolve to incorporate aspects of the conspecific courtship song into its sexual aesthetics. Both we and female canaries find the male canary's song enchanting, but we can be sure that robins and turkeys and certainly crickets and frogs find nothing sexy about those voices. The details of how the acoustic-aesthetics of choosers' brains become wired to favor their own singers varies. For example, as we discussed with túngara frogs, the frequencies to which their inner ears are tuned start the process; while in crickets the auditory nerves in the thorax are more responsive to the rhythm of the chirps; and in birds very little bias toward conspecific sounds arises in the periphery, such as the inner ear, but are resident in various areas of the brain. The brain has numerous ways to bias it to the sounds of its own species.

Once the brain is wired for its own species' song, some songs will better match the percept of what that species is supposed to sound like. There might be no differences among courters in their health, their resources, or their genes, but some might just happen to better match the resident aesthetics, sound a bit more like a canary is supposed to sound, for example, than do their sexual competitors. These are the courters who will get more mates, because—and only because—they are sexually beautiful.

Vocalizations do not only vary among species, but there can be substantial differences among populations of the same species. A friend of mine, Eddie Johnson, had a faculty position in Idaho, far from his birthplace in Brooklyn, New York. Despite having been out of Brooklyn for many years, Eddie's accent was straight out of the Bowery Boys; no one could have mistaken its origin for anywhere else. No one, that is, except a secretary in his department. While I was lecturing there, she told me how compassionate she thought the faculty were to hire someone with such a severe speech impediment; in her mind, the English language seemed to have only one dialect, and it was hers. In *Pygmalion*, and its adaptation *My Fair Lady*, Dr. Henry Higgins is able to place any English person's origins with amazing accuracy, based only on his or her accent or dialect.

Differences in dialect are also known to occur in animals, especially songbirds. The canaries and zebra finches discussed previously are only two of the more than five thousand species of songbirds. Like frogs, different species of songbirds have different songs, and unlike frogs, songbirds learn their songs from their fathers or neighbors early in life. Some songbirds never learn another note after that, while others are able to continue to expand their repertoires year after year. Nobody, neither bird nor beast, learns things perfectly. It is quite common for a male to grow up to sing slightly differently from his father. Over generations these differences accumulate, and eventually one population of White-crowned Sparrow, for example, will sound a good bit different from another population. They have different dialects. Does this matter? It seems that it does, because not only do males learn what songs to sing while in the nest, but females learn what songs are attractive. A number of studies, especially those with White-crowned Sparrows, show that females prefer the songs of males of the local dialect.

There are many anecdotes of dialect preferences in humans, but it is not always a preference for the local dialect. I know many women from the southern United States who find the *Sopranos*-like dialect of certain New Jerseyans quite unappealing, and I know a lot of Jersey boys who melt when they hear the drawl of a southern belle. In birds and humans, dialect preferences need not be tied into any utilitarian benefit for the chooser, even though we can imagine that there could be benefits. For example, some have suggested that a bird's dialect indicates it is best adapted to local habitats, so females that reside in these habitats should prefer those males. The suggestion is logical, but there are scant data to show that it is biological, that this really happens in nature. It might be merely that choosers like what they find familiar, or in some cases what they find exotic.

Our dialects can hint not only at where we are from geographically but also where we are perched on the social ladder. Consider the following sentences in which only a single word, the one in parentheses versus the preceding word, varies:

> Jack was waiting (wading) for a bus in front of his school.
> Jack had been playing (plain) sports all day and was very tired.
> He was worried about falling (fallen) asleep on the bus and missing his stop.

In each of these sentences the homophone in parentheses is typically associated with lower socioeconomic status. When young women from Ontario heard these sentences read by similarly aged men from Scotland, the women were acutely aware of social status and found the more "proper" vernacular to be the more attractive one. When women choose partners, they want resources; and, Jillian O'Connor and her colleagues from McMaster University and MIT argued, what is a better predictor of resources than diction?[5]

* * *

Certainly for humans, and also for most other animals, the most common mate choice decision is not the one between species or between courters with different dialects but the decision between courters of the same species in the same population. Females are not easily seduced, and males must be quite persistent in their vocal advertisements if they hope to secure a mate. Many songbirds, crickets, and frogs will call thousands of time a day in an attempt to persuade and seduce their females. These behaviors are costly: the rate of oxygen consumption increases substantially, as does lactic acid in the muscles, when calling. In some species, males lose weight, stress hormones increase, and testosterone decreases after days of calling, which then forces males to take a break for a few days to replenish their energy reserves. This energetic cost of calling should act as a filter to eliminate the more sickly and least healthy males from participating in the sexual marketplace.

There is an additional cost of broadcasting sounds for sex, namely, eavesdroppers. Eavesdroppers abound, and listening in on the conversation of others is a common means by which predators and parasites find a meal or a host. I have already discussed one expert eavesdropper, the frog-eating bat. The eavesdropper par excellence, however, is a parasitic fly called *Ormia*. These flies have evolved an intricate ear, unique among all other insects, which allows them to hear the chirps of crickets. Females use the crickets as a host for their developing larvae. A female alights on a calling male, and her larvae crawl off. They then begin to burrow inside the male. They eat him from the inside out as they develop, eventually killing him. Diabolically, they first feast on his calling muscles, which mute his song. This keeps him from inadvertently attracting more *Ormia* with competing larvae. *Ormia* invaded Hawaii about one hundred years ago, and the local crickets are paying a price. Marlene Zuk and her coworkers showed that parasitism has become so high on the Hawaiian island of Kauai that crickets have evolved the ultimate adaptation to thwart them—silence.[6] Male crickets make sound by rubbing their wings together; as the file on one wing is moved across the scraper on the other wing, it "chirps." The mutation that prohibits calling in the crickets on Kauai is one that changes the shape of the wing; "flat-wing" males are unable to call, and they have to intercept females passing through the area if they are to mate. Interestingly, the silent, flat-wing mutant recently appeared on the nearby island of Oahu. It seemed most likely that the mutant crickets had done some island hopping, but this is not the case. Nathan Bailey and his colleagues have shown that the genetic mutations that lead to flat wings are different on the two islands.[7]

The ability to avoid predators and parasites could act as another filter that restricts the sexual marketplace to the healthier, or perhaps the stealthier, males in the population. In most cases, we do not know if the males who avoid predators and parasites while calling have good genes or good luck. In the case of the Hawaiian crickets, however, we know that it is a lucky mutation that takes them out of the calling game.

There are many types of information about a courter that a chooser can garner from listening to how he sounds. Those vocal folds we all flap when we talk are not all the same size, and in mating, size often matters. In general, the larger the vocal folds, the slower they vibrate and thus the lower frequency of sound they produce. Not surprisingly, larger people have larger vocal folds and thus lower-pitch voices; controlling for body size differences between the genders, men still have lower voices than women. This is because testosterone causes an increase in the mass of the vocal folds. In the same study of preference for sociolinguistic differences, O'Connor and her team also showed that women preferred lower-pitched, or as they put it, more masculine voices.[8] Their explanation is that higher testosterone levels can indicate good health, and women want healthier males. This preference could be a direct benefit to both the woman and any children she and her baritone-beau produce, as he is likely to be around longer to care for them all. It is also possible that this type of preference provides indirect genetic benefits if his greater health will be passed down to his offspring. Therefore, choosing more masculine voices could also be choosing genes that are good for survival.

* * *

In the previous chapter, I reviewed numerous ways in which visual displays of courters have evolved to exploit some of the fundamental processes of the visual brain. Given the way the auditory brain works, there are several strategies that courters can adopt to make themselves more attractive to choosers. Also in chapter 4, I talked about how important it is for courters to be seen; it is equally important for them to be heard.

I live in Austin, Texas, the self-proclaimed "live music capital of the world." A lot of that music is performed outside, and much of it can be heard from great distances. Standing a reasonable distance in front of a stage, I can hear the crisp, rapid pulses of a saxophone and the high pitch of a fiddle as well as the slower beat of the drum and lower pitch of a bass guitar. But as I walk away, eventually those crisp pulses all blend together into one continuous sound, and the high notes of the fiddle dissipate into the atmosphere. If I keep going farther, all that my ears can pick up are the banging of the drums and the booming of the bass. This is because not all sounds transmit as efficiently over distance and across environments. Courters would benefit to take note of a few generalities about sound transmission: the faster the rate of separate sounds, be they notes of a sax or syllables of a sparrow, the more the pulses degrade with distance; the higher the frequency of sound, the more amplitude it loses with distance; and the denser the habitat, such as forests as opposed to fields, the greater the loss of pulse structure and high frequencies.

Are courters, or at least their genes, heeding these general principles of the physics of sound? The courtship calls and songs of many animals are loud screams for attention rather than tender whispers of intimacy. The larger the distance over which a courter is heard, the bigger its audience of potential mates, and in a number of cases animals have evolved those sounds that increase the size of their sexual audience. The ornithologist Eugene Morton started the discipline of habitat acoustics when he surveyed the songs of more than a hundred species of birds in the forests and fields of Panama and showed that birdcalls in the field had higher frequencies and faster pulse rates than those in the forest.[9] These birds have evolved to be heard and to be heard more clearly over longer distances by employing those sounds that do best in their own habitats: high frequency and fast pulses in the field, and low frequency tones and whistles in the forest. Great Tits in Europe as well as cricket frogs in Texas even show acoustic adaptations among different populations within the species.[10] The Great Tits, for example, use faster pulse rates that resemble Morse code when singing in the fields of Morocco but resort to more tonal singing when in the forests of England. When we hear birds, crickets, and frogs singing in the far distance, it is no accident; their songs have evolved in response to some basic principles of physics to travel far, and we just happen to be around to hear them.

These species have had plenty of time to evolve courtship songs that are a good match for where they sing. But what happens to birds that are suddenly stuck in urban environments with their onslaught of anthropogenic noise? There probably has not been enough time to evolve noise-proof songs, but changes in behavior are not restricted to the humdrum rate of mutation and selection. Many behaviors are flexible and serve as rapid-response teams that keep the organism going until evolution catches up.

Hans Slabbekoorn and his colleagues in the Netherlands have shown that songbirds have the know-how and the behavioral flexibility to take matters into their own hands, or at least their own voices, and not wait around for genes to mutate. When the researchers compared the songs of Great Tits in urban areas compared with tits in more rural ones, they found that the city birds used higher-pitched syllables that boosted their songs above the frequency band of the city sounds. One of his students, Wouter Halfwerk, then showed that males who sang these higher-pitched songs were more likely to be chosen as mates, probably because it was easier for the females to detect those signals against the noise.[11]

From a courter's perspective, noise is anything that interferes with his signal. The biggest source of noise for most courters is not the wind, the sounds of other species, or even the din of the city, but the guy next door. A courter needs to stand out against this noise so that he is the one heard, the one the chooser notices. One way to do this is to raise your voice as the surrounding noise levels increase. This is called the Lombard Effect when it occurs in response to what we typically consider noise, like the wind or the city, but courters also raise their voices when competing with other courters. Calling louder is not the only solution to standing out. As noted in chapter 2, túngara frogs add more chucks to their calls when competing with other males, despite the costs of increased predation risk. Other animals call faster, for a longer time, or add more notes. There are many solutions to being heard against "social noise" and it seems that animals have figured out most of them.

There is a lot going on in the world around us, and we are constantly being bombarded with stimulation. As mentioned in chapter 3, we tend to ignore repeated stimulation and return our attention when something changes. We habituate to the same old–same old, and dishabituate when there is something new and worthy of our attention. This is another basic principle that courters use in designing their sexual signals.

One of the striking features of songbirds is they often have quite large song repertoires. Nightingales, for example, can produce more than 150 song types. Numerous mimetic birds, like mockingbirds and birds of paradise, will imitate songs of other species, as well as pianos and even lawn mowers, to extend their repertoire size. In most cases studied, females find these larger repertoires more attractive than smaller ones. Why so?

More than a half-century ago, Charles Hartshorne, a well-known philosopher of "process theology" and a bird enthusiast, suggested the "monotony-threshold hypothesis" to explain why birds evolved large song repertoires.[12] Hartshorne thought complex birdsong was better at holding the attention of neighbors who might otherwise intrude on one's territory. This idea was seconded by the famed animal behaviorist Peter Marler, quoted in the epigraph to this chapter, who also argued that variety in songs evolved not to expand songs' meaning but to maintain interest. Hartshorne's idea has some support from the behavior of female birds, their auditory neurons, and their genes.

The ornithologist William Searcy showed that a female grackle might be initially attracted by a male's song, but she loses interest as the male repeats the same song syllable over and over again.[13] If the syllable changes, however, the female's libidinous desires return, and she responds with courtship solicitation displays, in effect signaling "come on, let's get it on!" A parallel phenomenon occurs at the level of the neuron, as shown by the neurogeneticist David Clayton.[14] When a zebra finch hears the same song syllable repeatedly, her auditory neurons stop responding: they habituate. If the syllable changes, the neuron responds again; it is released from habituation. The female's boredom and release from it are not restricted only to her behavior and neurons, but also influence her DNA. Expression of the gene *zenk*, which indicates signal saliency, is suppressed in response to the same repeated syllable and enhanced when a new syllable is experienced. From these studies, we have gained some idea about why female songbirds find more loquacious males more beautiful—they are less boring.

For any sexually selected trait, we can ask why males don't always evolve the most beautiful, and the answer is usually that they cannot afford the costs. Túngara frogs can chuck more than they do, but they are held at bay by frog-eating bats; crickets could chirp almost endlessly, but if they do so, they are more likely to become the home and meal for parasitic flies. What keeps male songbirds from adding an endless variety of notes? Elizabeth and Scott MacDougall-Shackleton show that in Song Sparrows, repertoire size also acts as a hurdle to the less healthy.[15]

Songbirds have pretty special brains to produce these wonderful melodies that keep us and their females so entranced. Researchers know a lot about how the bird brain generates song. One of the most important areas is called the HVc, for "higher vocal control center." In general, the HVc is larger in species with larger song repertoires; it is larger in males than in females; and the difference in the size of the HVc between the sexes varies as a function of the size of the song repertoires of males and females. The MacDougall-Shackletons found that in Song Sparrows, males with large repertoires not only had larger HVcs, but they were in better body condition. They also exhibited less physiological stress and a more robust immune system compared with their smaller-brained and less loquacious brethren. This could mean that males with more songs might be better fathers since they are physically more fit, a direct benefit for the choosing female because she might then be able to produce more young. If there are genetic differences, rather than developmental differences, that promote larger repertoires, bigger brains, and healthier offspring, this would pass down some indirect genetic advantages to their offspring as well.

* * *

Although choosers should be concerned about the quality of their mates, whether it be their species membership or their general health, matings are only worthwhile if they produce offspring. As the abundance of fertility clinics attests to, having sex and reproducing are not the same thing. For reproduction to take place, the male and female need to be on the same page physiologically. Males are usually ready to go, as shown in chapter 1, but the development of eggs is more complicated than that of sperm; thus a female's fertile period is shorter than the male's. But the male can tweak the female's hormones and hasten her horniness. But the female's internal reproductive physiology is discerning, and males have to do things just right.

Ringdoves (*Streptopelia risoria*) are close relatives to the common pigeons that populate many cities around the world. If you have spent any time in a big city, especially New York or Venice, and watched the pigeons instead of just feeding them or shooing them away, you might have noticed a male making cooing sounds and vibrating his vocal pouch while he strutted around a female. Even if you watched closely, however, you probably would not see them mate. That takes place only after a male entertains the female for days, and when it does finally occur, it is over in a jiffy. For most birds, the sex act consists of a mere "cloacal kiss," as males of most species lack anything like a penis.

Researchers have studied the ringdove to understand how this cooing gets inside the female's head and then straight to her hormones. The details have emerged over many years of research initiated by the late Danny Lehrman, one of the best-known comparative psychologists of his time, who founded the Institute of Animal Behavior at Rutgers University in Newark, New Jersey, a city which seems to have enough pigeons to feed the world. Lehrman showed that courtship in ringdoves is not restricted to males displaying to females but instead is a series of detailed interactions between the sexes.[16] The initiation of courtship is what we are seeing on the city streets. If a male is sexually receptive himself, specifically if his testosterone level is above some threshold, he begins to court. A male bows his head and begins to coo as he struts around the female in a display aptly termed the *bow-coo*. The male's cooing influences the female's sex hormones, causing her estrogen levels to escalate, and she now calls back, joining in a duet with him that affects him in two ways—his testosterone increases, and he courts the female even more vigorously. At some point, the female's estrogen levels begin to drop, and the hormone that modulates parental behavior, prolactin, begins to rise; the male and female then start to make a nest together. Then, and only then, the male and the female kiss, not with their mouths but with their cloacae. With the female's cooperation, the male climbs on top of her; they align the openings of their sexual organs; and the male expels of bit of sperm inside the female. Now that his deed is done, the male's testosterone drops off and his prolactin increases, which primes him for his paternal duties, as he shares in both incubating the eggs and feeding the nestlings. The entire process, from first bow-coo to the final kiss, takes days.

Some interesting details were added to this already quite complete story by one of Lehrman's protégés, Mae Cheng.[17] It was thought that early in courtship the male's call caused the female to call, and that her act of calling caused her estrogen levels to increase. Instead, Cheng discovered, it is the female *hearing* herself call that stimulates her own hormone levels. And, what is especially surprising, the female only responds to the male if he is looking at her. At least in this species, a male's wandering eyes can doom a relationship.

In the ringdoves, we see how the female sexual brain is hooked up to her sexual physiology. Her decision to reciprocate a male's sexual interest both influences and is influenced by her sexual hormones. The male must push all the right buttons, keep pushing them for days on end, and give the female his undivided attention to get her physiology to communicate to her brain that this male is sufficiently attractive to be rewarded with sex. Ringdoves are not unusual in using courtship to tweak the hormones of those desired for sex. Singing in songbirds, chirping in crickets, calling in frogs, and roaring in red deer all influence female hormone levels, bringing them into the state in which they desire sex.

As I have been emphasizing throughout, sex is more likely to happen when one sex finds the other sexually beautiful. As just explained, it is also more likely to happen when both sexes are in a physiological state for sex. In chapter 3, I showed how liking sex is not the same as wanting sex. Similarly, reproductive hormones that cause females to ovulate, build a nest, clean the house, and tend to their offspring are not the ones that drive the desire for sex. That happens in the reward system, and recent studies of birds show how song seems to link up the reproductive system with the desire to reproduce.

Donna Maney and her colleagues have been exploring this link in White-crowned Sparrows.[18] We already know that in sparrows, just as in ringdoves and other songbirds, male song influences the level of reproductive hormones such as estradiol and, by extension, female physiological readiness to mate. These researchers also examined areas in the reward system, specifically the nucleus accumbens and the ventral striatum, which are where norepinephrine and dopamine are released and where "liking" is coupled to "wanting"—in this case, where the pleasure of hearing a male's song is linked with the sexual desire for it. They found that overall gene activity in these areas increased when females were exposed to courtship song. But not at all times . . .

The context in which sexual signals are produced and received can influence their meaning. Season and sex are two such contexts. The song of the White-crowned Sparrow serves several functions, and one is advertising his territory to other males. Males assign a negative salience to songs of other males and respond to them aggressively rather than sexually. Male White-crowned Sparrows sing in the breeding season, in the spring, but they also sing in the winter, out of the breeding season. Females are attracted to singing males in the breeding season but will attack those same males when the males are singing out of the breeding season. The difference in the female's response between the seasons is based on her estrogen levels. Courtship song only turns on the reward system when her estradiol levels are high. They only trigger sexual desire when she is reproductively ready. Thus males can use sound to tweak not only the female's reproductive readiness but also her sexual desires, her liking as well as her wanting. But only if all the stars are aligned. Song only triggers liking and wanting if the female is hormonally ready for reproduction. Liking sex, having sex, and wanting sex are not the same as reproducing, and all of these systems have to be aligned for choosers to desire their courters. In most animals, the act of sex is inextricably linked to the function of reproduction, and it is not surprising that some sexual traits evolved to trigger both of these functions.

* * *

Two major themes of this book are that the brain has other things on its mind besides sex, and that these other brain functions can influence which stimuli are perceived as sexually attractive. Courters can also trigger behavioral responses in choosers that have nothing to do with sex but that up the courter's chances of having it. In chapter 4, we saw how courters use certain visual displays to exploit the chooser's need for food and their desire not to become food. Acoustic courters can be just as deceitful.

Animal hearing sensitivity can evolve for hearing sounds related to sex or for other functions. The ears of crickets, for example, have auditory neurons that are tuned either to the sonic frequencies of mating calls or the ultrasonic frequencies of bat predators. Ron Hoy, an expert on insect hearing, and his colleagues showed that crickets reduce most of the variation in what they hear into these two categories, depending on whether the sounds they hear are below or above 16,000 Hz—females approach sounds in the lower-frequency category and flee from those in the higher-frequency category.[19]

Moths are another groups of animals that use their hearing to avoid predators. Many moths have evolved ears that hear bats and organs that produce ultrasonic calls to jam the bat's echolocation calls. We are used to seeing moths at night, and most of them are in fact nocturnal. There are some, however, that have abandoned the night for the day, where birds rather than bats become their main predator. Even though the danger from bats is gone, these diurnal moths have retained their anti-bat weapons; they can still hear and call in the ultrasonic. Not to be wasteful, selection has co-opted these anti-predator adaptations to be used for courtship; ultrasonic clicks have now moved from the moth's arsenal of defense to its arsenal of courtship. Even some moths that are nocturnal recruit their acoustic bat defenses for courtship. The Asian corn borer is one of these moths.

The Asian corn borer is one of the worst pests in all of Asia, causing millions of dollars of damage and sometimes totally devastating corn crops. The female corn borer lays a couple of hundred eggs on a corn stalk, and the larvae bore into and consume pretty much every part of the plant. The eggs are often infected by bacteria, which only makes them more of a pest. The bacteria feminizes males, resulting in mostly female broods, which then enhances the growth and spread of the pests throughout agricultural fields in that part of the world. This moth also calls during courtship. The male rubs his two wings together to produce ultrasonic calls, sounds that initially evolved to battle bats but are now used to signal sex.

Not all the bat weapons the moths evolved are morphological; some are behavioral. Besides hearing bats and jamming the bat's echolocation, some moths go into steep, erratic dives or just freeze when they hear bat calls. The corn borers will freeze in response to bat echolocation calls, and the males have exploited this response in their females. Ryo Nakano, a biologist at the University of Tokyo, and his group have shown that males produce a low-amplitude call when engaged in courtship with females. The call is very similar to a feeding buzz of a bat, the rapidly pulsed series of echolocation calls the bat makes when it is rapidly closing in on its target for the kill. Nakano showed that if corn borer males are muted or females deafened, then courtship is usually not successful, but if males call and females listen, then sex is almost guaranteed to follow. The reason that this call is so successful is because females respond to the call as if it were made by a bat rather than a male moth. She freezes in fear, and while she is in this paralytic state, there is little resistance to the male mating with her.[20] Jim Morrison of *The Doors* said "sex is full of lies"; we can add that in this case those lies are entangled with fear.

The male corn borer moths mimic predators searching for food in order to trick their females. Heather Proctor found that male water mites mimic food to get a mate. Like other hearing animals, water mites are very sensitive to vibrations in the world around them, but these vibrations are in the water rather than the air. This is a good thing, since one of their favorite foods are copepods, who set up a very characteristic pattern of vibration as they scoot across the water's surface. Males have evolved to mimic this vibration pattern to lure females who do not realize that the source of the signal is a male until she grabs him, at which time the male begins to vigorously court her. If hunger motivates the female's approach to the male, Proctor predicted, in nature hungry females should be the ones more likely to be duped by the male food-mimics and thus these females should be the ones most likely to mate. She conducted a clever experiment using two groups of females: mites in one group had not been fed for several days, and mites in the other group were able to dine to their hearts' delight. Males were then added to each group and the number of matings counted. Females who were hungry for food were the ones more likely to mate, as predicted; their higher rate of mating was entirely linked to their desire for food, which then resulted in their being fooled into sex.[21] We see that not only can courtship signals evolve to match sensory, perceptual, and cognitive biases, it can also exploit behavioral responses that have nothing to do with sex.

* * *

The renegade-model Cameron Russell and numerous animals have shown us that we are not stuck with the looks with which we are born. Neither are we stuck with the songs we can sing. If we have some help, some imagination, we can improve on what we have; we can extend our acoustic phenotypes.

On one trip I took with Stan Rand, we searched for a close relative of the túngara frog in coastal Peru. This region of northwestern Peru, which includes the Sechura Desert, seemed an inhospitable place for both humans and frogs. We drove for miles seeing little vegetation and no water. The prehistorical inhabitants of this area, the Chimor, who grew out of the Moche civilization, survived on both agriculture and the sea from about AD 900 to 1470, at which time they met their demise at the hand of the expanding Inca Empire. The Chimor had a predilection for marine mammal meat, and some of their drawings show nets for capturing sea monsters.

The most spectacular of the ruins the Chimor left behind is Chan Chan, a UNESCO World Heritage site and at one time the largest adobe city in the world. Stan and I wandered into an amphitheater that had been used for ceremonies, probably both religious and political. Whether priest or politician, we can be sure that whomever was speaking wanted to be heard. Of course, boom boxes were off in the distant future, and there was no ability to electrically amplify sounds. But there was the science of acoustics, and the Chimor architects applied some of its basic principles to designing this amphitheater, shaping it to project the voice of a speaker to an attending crowd. When a speaker stood in a particular place his voice would be resonated, its amplitude enhanced, and his words of wisdom would boom across the crowd. We tried it for ourselves. Stan took the place of the speaker, and I moved back into the area where the crowd would have assembled. Stan imitated a túngara frog call, and it reached me with an amplitude and richness that the real frogs could never match. I almost dropped on all fours and hopped to Stan! The Greeks, Romans, and other groups in the Western Hemisphere also mastered this trick in acoustic engineering, and the practice of using our environment to enrich our voices never died out.

Growing up in the Bronx in the 1950s in my pre-teens, my friends and I often came across groups of young men congregating in alleyways and building foyers, combing their greasy hair straight back to look like Elvis; yes, they were greasers. Although the Bronx was certainly not crime-free in those days, these guys were not shooting up or even drinking down. They were singing, usually in syncopated harmony, songs by the likes of the Everly Brothers, Buddy Holly, and the Platters. They sought out these small enclosures because of the impressive effects on their singing. We would watch and listen from a distance. The songsters usually ignored and tolerated us. We never heard a sound this rich and resonant from our pocket-sized, Japanese transistor radios. These guys were pretty good at enhancing their voices, at extending their phenotypes. Like the Chimor, the greasers used the walls around them to aid their natural voices.

As noted earlier in this chapter, when courters use sound, they usually want to be heard over the greatest area possible so as to reach the biggest audience of choosers. When choosers can hear calls of more than one courter, they usually prefer the loudest of the calls. Like the greasers in the allies of the Bronx and the Chimor in the amphitheaters of Peru, some animals have figured out how to enhance their courtship sounds in this way.

Many species come with built-in traits that enhance the amplitude of their calls: frogs and howler monkeys have large vocal sacs; cicadas have resonators on their body walls; and whales have resonators in their heads that ultimately help their sounds come out louder and travel farther. Others engineer the environment to match their voice or vary their voice to match the environment.

Sound frequency and sound wavelength are negatively related to one another. Higher-frequency sounds have shorter wavelengths, and lower-frequency sounds have longer ones. If you play sound into a tube, the sounds that come out louder are those whose wavelengths match the length of the tube. In woodwinds, such as flutes, sound is produced by blowing air across a narrow hole at one end of the instrument, which then causes the air column within the flute to vibrate. The frequency of the vibration, what we perceive as pitch, is determined by the length of the air column: the longer the flute, the lower the pitch. But flutes are not stuck with the pitch determined by their length. Musicians can extend the flute's sounds, its acoustic phenotype, by changing its effective length with a mere touch of the finger. By opening and closing holes on the instrument, its effective length changes. When all holes are closed, the effective length is the longest and the pitch the lowest, whereas opening various holes decreases the effective length and increases the pitch. Humans have applied basic acoustics to engineer a variety of pleasing sounds we call music, and animals have done the same to enhance their own music.

Some crickets and frogs in Australia, like the Chimor and the greasers, use chambers to resonate their calls. They call from inside burrows in the soil that, depending on the species, they either construct at a length that best matches the wavelengths of their mating calls, or they position themselves in an abandoned burrow at the distance from the opening that gives them the best resonance.

The herpetologist Jianguo Cui and his colleagues have shown that the Emei music frog in China goes one step further and uses the interaction between its call and its environment to advertise its real estate to females. This is how it works: a male calls from inside a nest he constructs to hold a female's eggs. The calls that emanate from inside the nest have more energy in lower frequencies and longer notes compared with the same calls that are broadcast from outside the nest; the effect on the inside-nest calls is mostly due to the size of the burrow entrance and the depth of the burrow leading to the nest. When females were given a choice between the same call broadcast from inside and outside of the nest, a vast majority preferred the longer, more resonant call from inside. They prefer the males with better "real estate."[22] These vocalists engineer their environment to match the call, but a frog in Borneo does just the opposite: it manipulates its call to match its environment.

The Bornean tree-hole frog, as its eponymous moniker suggests, is found in the forests of Borneo, where he calls from tree holes. The tree holes differ in size, and the size of the inner air-filled cavity also varies with the amount of water that has accumulated in the tree hole. The wavelengths that are best resonated depend on the size of the empty cavity. There is not much this frog can do to change the cavity size; he can't carve out the tree, truck in rainwater, or siphon it out. Being stuck with his tree hole and the amount of water in it, this frog's solution is to adjust his calls' frequencies and wavelengths to best match the cavity's size so it increases the calls' amplitude to reach as many females as possible. How do we know this?

In a very clever experiment, two researchers, Björn Lardner and Maklarin bin Lakim, placed frogs in artificial cavities partially filled with water. The researchers recorded the frogs' calls as they slowly drained water from the faux tree hole. As the water drained, the cavity space became larger, making longer and longer wavelengths the ones that would be best resonated. As all this was happening, the frogs actively changed the wavelength of their calls to match the cavity's resonance.[23] So as with the Chimor and the Bronx greasers, when animals have something important to say, they use all kinds of tricks to be sure everyone hears it.

When we think of animal sounds, we usually think of voices. There are lots of other ways to make sounds, however. Crickets, whose names come from the French *criquer*, meaning "little creaker," rub a file on one wing along a scraper on the other to produce the chirps so characteristic of summer nights in the temperate zone. Cicadas take a different approach and vibrate a drumlike tymbal organ on their body wall. Toad fish shake their sonic muscles two hundred times per second, the fastest muscle movement of any vertebrate, to make their swim bladders hum, which is a great alternative if you can't sing. In all of these examples, as well as our own voices, to make sound you have to vibrate something. I once came across what is probably the most direct way to produce sound one sunny afternoon in the El Duque Reserve in Amazonian Brazil. Although there was not a cloud in the sky, I could swear I heard rain dropping on the dry leaf litter on the forest's floor. I kept peering upward in vain for the rain that was responsible for this sound and saw nothing; when I finally looked down, I saw hundreds of ants scurrying around. When I got on my hands and knees for a closer look, I saw that they were all banging their heads into the ground; there were so many of them and they were banging their heads so hard that they really sounded like rain. Eventually, the ants settled down and their sound ceased. What was this all about? I found their nest, stuck a stick inside, and out came the ants. The sounds of banging heads once again filled the forest. I thought I had discovered a true oddity of nature, but a little research when I got home revealed that although an odd behavior, it is a well-known one in a number of ants and termites. The function of the sound is to warn nest-mates of predators, or of field biologists molesting their nest. There are a lot of ways to make sounds besides with a voice, just vibrate something. As we will now see, voiced animals can be very creative about adding to the sounds that come out of their mouth.

Hummingbirds are pretty amazing animals. They can beat their wings up to forty times per second, and while doing so they can stay virtually motionless in space while they delicately insert their extra-long beaks into the nectary of a flower. Like songbirds and parrots, they also learn their songs. For me, the most impressive characteristic of these small, seemingly fragile birds is the aerial courtship dive. In both Anna's and Costa's Hummingbirds, when a female enters a male's territory, he hovers and sings in front of her as he flares the patch of glistening feathers on his throat. If the female doesn't spook, he then treats her with some amazing aerial acrobatics. He flies thirty meters into the air and then launches himself like a dive bomber and hurtles toward the object of his sexual desire; he might repeat this dive up to twenty times. To ensure that he has the female's attention, he punctuates his dive with a loud buzz. This sound was long assumed to be part of the song until Chris Clark, then a graduate student at the Museum of Vertebrate Zoology of the University of California, Berkeley, showed that these sounds are produced as air passes over and vibrates the bird's outer tail feathers. He examined a dozen or so species, including the Anna's and Costa's Hummingbirds and their closest relatives, and found that most of the species have aerial courtship that always includes tail buzzes. But in a smaller subset of these species, males also add sounds produced the old-fashioned way—they sing them. The vocalized sound and the tail buzzes are so similar that it was long assumed they were both part of the song. Because the buzzes evolved before the songs, Clark concluded that the similarity between the two sounds must have resulted from the song evolving to mimic the tail buzzes.[24] Even though hummingbirds are vocal animals, in this case the voice plays second fiddle to the tail. Other birds literally play fiddles—or more appropriately, violins—to enhance their courtship.

Manakins are the real pranksters of avian courtship. Unlike songbirds and hummingbirds, they lack the ability to learn song, but as a group they experiment with a variety of sounds. The crown jewel of manakin courtship to my eye, or more accurately to my ear, is the Club-winged Manakin. In my classes I show students a video of one of these birds from a study by Kim Bostwick of the Laboratory of Ornithology at Cornell University, which features a close-up of a brown-chested, red-capped male on a perch. When he sings, he bends forward and quickly erects his dark wings with their light chevron markings, and he continues to hold them upright while producing a rapidly pulsing sound similar to a violin. I ask the students to explain what is happening, and each year I get the same answers: the male is singing as he bends forward, and he raises his wings as a visual cue to supplement the sound. We then look at the video in slow motion. The more observant students notice that the beak is closed when the sound is produced; this is normal for a frog, but humans and birds open their mouths when they vocalize. The most observant students notice a very slight and rapid back-and-forth movement of the erect wings.

Bostwick realized that these wing movements are responsible for these sounds. Aiding her laserlike observational powers with real lasers, she showed that males are vibrating their wings one hundred times per second, more than twice the rate at which hummingbirds do. Their wings have a morphology that shares some of the basic properties of a violin. Each wing has a specialized feather with a series of ridges and another feather with a stiff up-curved tip. When a male raises his wings and vibrates them against one another, the tip on one wing strikes against the ridges on the other wing and . . . viola, the forest is alive with the sounds of violins![25] This is just one more example of the creative extent to which animals' brains, morphologies, and behaviors conspire to mold their sounds of sex to please the senses of their courters. But nobody engineers sound better than we do.

* * *

I am in Edinburgh as I finish up this chapter, a World Heritage City where the Scottish culture flows more abundantly than the 250 flavors of whisky at its famed Albanach Bar. It is a crisp, sunny, spring day; the grass is green; the flowers are colorful; and there is a kilted bagpiper every few blocks. Standing at a corner waiting for the traffic light to change, the bagpipes command me to move. The *Sturm und Drang* is unmistakable, the pitch, the rhythm, the phrasing, everything about this music is martial. Although the piping did not quite make me want to fight or become aggressive, the music had an effect—it was calling me to at least move if not march. All of a sudden, the red light seemed interminable.

Animal songs and calls are not evolutionary precursors of human music. They do, however, share many similarities. Both are primarily social behaviors. They can establish social bonds, reduce and initiate conflict, and, of greatest concern to us, both can be intricately entwined with courtship and sex. The effectiveness of animal songs and human music derive from the fact that both influence the affective or emotional state of the chooser through the structure of the sound. This is unlike language, in which the structure of the sound is usually arbitrary relative to the meaning assigned to it; in courtship and music the details of the sounds themselves are both the message and the meaning. Given that this is the case, we expect to find some generalities in how acoustic courtship signals and music elicit some similar emotions. And we do.

Earlier, I discussed Eugene Morton's discovery that the acoustic properties of the local environment influence the evolution of the structure of birdsong. He also proposed a set of "motivation-structural rules" to predict how different sounds of some birds and mammals elicit different emotional responses from their receivers.[26] There are eight specific rules, but Morton summarized them more briefly: "birds and mammals use harsh, relatively low-frequency sounds when hostile and higher-frequency, more pure tone–like sounds when frightened, appeasing, or approaching in a friendly manner."

We know from our own experience that different sounds better fit different situations. The best understood example might be "motherese" or baby talk, a cooing pattern of higher frequency speech that is cross-culturally used when interacting with infants; not only mothers talk to babies like this. When we want to soothe someone, child or adult, we tend to use quiet sounds that are long and tonal with long onset and offset times, that is, the amplitude slowly increases in the beginning and then slowly decreases at the end to avoid startling the receiver— "oooooohhh." In contrast to this type of soothing speech, if we are in an argument or a fight, we raise our voice and employ sounds that are short, harsh, and have quick onset and offsets. In anger we could protest, "screeeeeeeew yoooooooouuu" in a tonal voice, but instead we ejaculate, "Screw yoU!" in a manner that sounds as harsh as its meaning.

Our own rules about the functions of different sound structures are also applied to how we vocally interact with animals. Whether tending camels, walking our dog, or riding a horse, we use short, harsh, clicking sounds to initiate locomotion and more tonal, drawn-out sounds to command the animal to stop. When we do this, it might seem as if we are imposing our own structure-function rules on animals, but instead, we are using sounds that match their own structure-function rules, which just happen to be similar to ours. Patricia McConnell is a world-renowned dog trainer and author whose titles include *Feisty Fido*, *The Other End of the Leash*, and *The Cautious Canine*. As a graduate student, McConnell delved deeply into the command sounds used by trainers. In one experiment, naive domestic dogs were first trained to respond to the types of sounds that typically signaled "go" and "stop": four short notes with a rising fundamental frequency indicated go, and one long note with a descending fundamental frequency indicated stop. Another group was raised to stop when it heard the four short notes and to go when it heard the long note. Each group was then trained to learn the opposite association. Initially, both groups learned to associate go and stop with the acoustic signals with which they were trained. But when trained in the reversal learning trials, the group that was trained from the atypical association to the typical one learned faster than the group that first learned the typical association and then was asked to learn the atypical association.[27] This study, together with Morton's theory of motivation-structural rules, suggest that there might be some generality of structure and function over a wide range of species, including our own, because these particular sound structures interact with the receiver's neurobiology and psychology in similar ways.

Music is similar to animal signals in that it can elicit a variety of emotions. Here is part of a list from two Swedish psychologists, Patrik Juslin and Daniel Västfjäll, on the responses music evokes: *subjective feeling*, listeners report that they experience emotions while listening to music; *physiological reactions*, similar to those shown to other "emotional" stimuli, music induces changes in heart rate, skin temperature, electrodermal response, respiration, and hormone secretion; *brain activation*, responses to music involve regions of the brain implicated in emotional responses; *emotional expression*, music makes people cry, smile, laugh, and furrow their eyebrows; *action tendency*, music influences people's tendencies to help other people, to consume products, or to move.[28] Yes, just try listening to bagpipes when you are waiting for the light to turn green.

Different types of music evoke different emotions, although defining the features that cause each emotion can be quite complicated. One generally recognized connection is that different musical keys tend to evoke different emotions: songs in major keys tend to elicit happiness; those in minor keys are sad; and songs played with a blues scale are, well, bluesy. Centuries ago, Christian Schubart offered a detailed description of the emotional aspects of different keys in his *Ideen zu einer Aesthetik der Tonkunst*, which Rita Steblin translated in *A History of Key Characteristics in the 18th and Early 19th Centuries*. His descriptions are reminiscent of an overindulgent wine connoisseur. Here are just a few examples: "D Major, The key of triumph, of Hallelujahs, of war-cries, of victory-rejoicing. Thus, the inviting symphonies, the marches, holiday songs and heaven-rejoicing choruses are set in this key; D Minor, Melancholy womanliness, the spleen and humors brood; F♯ Minor, A gloomy key: it tugs at passion as a dog biting a dress. Resentment and discontent are its language; A♭ Major, Key of the grave. Death, grave, putrefaction, judgment, eternity lie in its radius." And finally, getting closer to our own concerns, "A Major, This key includes declarations of innocent love, satisfaction with one's state of affairs; hope of seeing one's beloved again when parting; youthful cheerfulness and trust in God; B♭ Major, Cheerful love, clear conscience, hope, aspiration for a better world."[29] As Schubart seems to imply, music can influence our sexual mood, and its actions can be as primal as the effect of the ringdove's bow-coo.

In an article published in the *Archives of Sexual Behavior*, David Barlow and his coworkers determined how music influences how we feel about sex. The experiment was simple enough. Give a man some happy or sad music, show him a porn flick, measure his erection, and ask how horny he is. The approach might sound a bit lowbrow, but not the music. They used samples from Mozart's *Eine kleine Nachtmusik* and Divertmento no. 136 to induce a positive mood and Albinoni's Adagio in G Minor and Barber's Adagio pour Cordes to induce a negative one. They do not report similar details about the porn videos. The results were what we might expect, penis tumescence and sexual arousal were both elevated when the subject was primed with music that induced a positive emotional state compared with the negative music.[30] It is no accident that romance in everyday life often includes a sound track.

Music also reaches deep into our brain, targeting the same reward areas that stimulate liking and wanting. Anne Blood and Robert Zatorre, both scientists at McGill University, conducted PET scans (positron emission tomography) on subjects who heard music that had a chilling effect on listeners. When the subject reported a sensation of "shivers down the spine" or "chills," the PET scans revealed increased blood flow to various regions of the mesolimbic reward system, showing activation of those same brain areas, such as the ventral striatum and nucleus accumbens, that were stimulated in White-crowned Sparrows and túngara frogs in response to their mating sounds, which they both liked and wanted.[31] Dan Levitin, also at McGill University and the author of the entertaining *This Is Your Brain on Music,* and his coworker Vinod Meno at Stanford University confirmed and extended these essential finding with fMRIs (functional magnetic resonance imaging), which provide more resolution than PET scans.[32] As noted in chapter 3, these reward areas are also exploited by addictive pleasures such as food, sex, drugs, and gambling. We now see that there is a linkage within the triad of sexual selection, dopamine, and "music" (including acoustic courtship), which nicely parallels the mantra of the 1960s—*sex, drugs, and rock and roll.*

Now that we have "seen it all" and "heard it all"—or at least seen and heard a lot about how sights and sounds interact with sex—we will move on to what might be our most primal of all senses and begin to understand the scent of sex.

SIX

||||||||||||||||||||||||||||||||||||||||||||||||||||||||||||||||||||||||||||||||||||||

The Aroma of Adulation

Smell is a potent wizard that transports you across thousands of
miles and all the years you have lived.   —*Helen Keller*

WE SEE, WE HEAR, AND WE SMELL. These three senses all channel stimulation from the world around us into our brains, where their information is merged and collated, some getting more attention than others, and where this information is then used to form our percepts about the world and to help us decide how to interact with it. All of these modalities can be important conduits into our sexual brain, and different animals usually depend more heavily on one modality than the others to identify potential mates and to learn something about them—their species, their sex, their health, and their readiness to mate. Some animals, including us, recruit all of these modalities to experience the sexual beauty of our mates. The sensations and the information we glean from each sense are different but often are complementary. Let's consider an example, unrelated to sex, of how these senses differ and how they can complement one another.

It has been dry for some time, as it often is here in Texas. We are in a drought that started six years ago and is still going strong. Lake Travis, the main source of water for the ever-growing city of Austin, is two-thirds dry; boat docks there have not felt the embrace of lake water for years. But sometimes I venture out onto my deck and I can immediately, even though it is not raining, sense that relief is on the way; the smell of rain fills the air. We have all sensed this heady fragrance but have probably thought little about its origin. The smell of rain is called *petrichor*, a word derived from the Greek *petra* (stone) and *ichor*, which, in Greek mythology, is the ethereal blood of the gods. Petrichor is produced by oils trapped in soil and stone, which become volatile when exposed to the moisture in the air that proceeds the falling rain. I am not the only one who gets excited by the smell of petrichor; during these same droughts cattle become restless when exposed to it.

The smell alerts me to the possibility of rain, but where is this potential rain? My sense of smell is of little help here, but suddenly I see a strike of lightning in the distant southwest. Now, I know that not only will it rain, but I know the direction of the oncoming storm. Two senses, two different types of information about the same phenomenon. But how far away is it? Do I need to quickly get under cover, or can I relax for a while? As noted in chapter 4, judging the distance of an object can be tricky. But five seconds after I see the lightning, I hear the thunder. They both occurred nearly simultaneously at the source of the storm. When lightning discharges, it heats the surrounding air to temperatures that exceed those of the sun; this causes the air to rapidly compress, which produces the initial loud crack of the thunder. The air then slowly expands, which gives us the rumbles that follow. Since light travels faster than sound, I saw the lightning before I heard the thunder. Light travels so fast, about three hundred thousand meters per second, that we can consider it instantaneous when we experience it within our own planet. (Once we breach our planet, it is a different story. It takes eight minutes for the sun's light to reach us, but still pretty quick for a 93 million–mile journey.) Sound, on the other hand, travels at a more leisurely pace of 330 meters per second. Because five seconds elapsed between when I saw the lightning and when I heard the thunder, I can reckon that the storm is 1,650 meters away, about a mile. To sum up this rather simple experience, I smelled it was going to rain, I saw the direction it was coming from, and I heard how far away it was. Who needs Doppler radar?

These three sensory modalities have both strong points and weak points when they are used in communication, including sexual communication. Visual signals result from reflecting light from the sun to a receiver, but if there is no light, there is no visual signal, so it only works well in the day. Visual communication is fast, and it gives very accurate information about the location of the source. You would never say about someone whom you could see standing on a corner, "I think she is somewhere in that direction." If you see her, you know exactly where she is, because seeing relies on a direct line of sight—if your friend ducks into a crowd, you have lost her, or at least you have lost visual contact with her. In acoustic communication, the sender creates its own energy, as noted in the preceding chapter, by vibrating body parts. Unlike visual communication, it is not restricted to daylight, nor does it require a direct line of sight. When your friend fades into the crowd, she can give a quick yell or whistle to help you to keep track of her whereabouts, although not as accurately as if you could see her. Both visual and acoustic signals are ephemeral, now you see them, now you hear them, but after the signal is produced, you don't experience it any more.

Olfactory communication shares fewer similarities with these other modalities. There is something about odors that makes them seem more primal than the color of someone's hair or the timbre of his voice. Consider this exchange between two characters in *Scent of a Woman*, in which the blind Frank, played by Al Pacino, is evidently charmed by the airline attendant, as he explains to his young companion:

FRANK: Where's Daphne? Let's get her down here.

CHARLIE: She's in the back.

FRANK: The tail's in the tail. Ooohhaa. Oh, but I still smell her. Women, what could you say? Who made 'em? God must have been a fucking genius. The hair. They say that the hair is everything, you know? Have you ever buried your nose in a mountain of curls, just wanted to go to sleep forever?

* * *

Olfactory communication only takes place when the molecules that constitute the odor make their way to the olfactory receptor cells of the receiver. In some cases, the travel is short and direct, as when dogs sniff butts. In other cases, the odors are deposited on an object, such as when dogs urinate on a tree or fire hydrant, and the receiver comes into contact with the stationary odor. In many cases, however, the odor molecules travel through the environment, riding the wind as they head for a chance encounter with a distant receiver. This is what happens when a female dog releases pheromones that indicate she is in "heat"; seemingly every male dog in the neighborhood gets the message that she is reproductively active and ready to mate . . . just listen to all that howling. Because the movement of molecules in the atmosphere can be tortuous, anything but a straight line, the directional information of odor cues is not very accurate. The only way to find the source of an odor is to follow its concentration gradient, which usually, but not always, increases the closer you get to the source. If you want to communicate your location, it is better to rely on a wave or whistle than your body odor. If your friend farts in a crowd, it is the sound more than the odor that will give away her location. Folks in the crowd will know the general direction from which the odor is coming, but not exactly who is responsible for it, as they exchange accusatory glances.

If you want to make a lasting impression, odor cues are the best. A dog who bares his teeth and growls at an intruding neighbor effectively defends his space, but when he leaves, those defense signals leave with him. A little bit of urine, however, goes a long way, or at least is present for a long time, and is a constant reminder to any intruder that it enters this area at its own risk.

Odors are important for different functions, but as with eyes and ears, there is only one channel that funnels all this information to the brain. The reception of odor is a bit simpler than the reception of light or sound, because the molecule that is the signal binds directly to the olfactory receptor. These olfactory receptor cells are in the noses of all vertebrates, the vomeronasal organs of some vertebrates, and in the antennae and feet of many insects. Some of these olfactory receptors are quite sensitive and can detect a few molecules of odor that have traveled kilometers from their source. The olfactory receptors on the antennae of the silkworm moth, one of the best odor detectors on the planet, captures 80 percent of the odor molecules that pass by, and the binding of a single odor molecule to a single receptor is sufficient to trigger the male's search for a mate. It is impossible to be more sensitive than that.

We need to pause for a bit of jargon. I have been using the terms *odors* and *pheromones* seemingly interchangeably, but they are not the same. Anything that gives us information is a cue, like the body odor of our lost friend. If that odor evolved specifically for communication, it is a signal. A pheromone is an odor-signal; it evolved to inform. I remember picking up a pamphlet on Telegraph Avenue in Berkeley, a home to the counterculture of the 1960s, many years ago when I was a postdoc. It was a diatribe against capitalism and rude people, titled *Flatulence as a Social Weapon*. Dialectics aside, it hopelessly confused the notions of odors and pheromones.

The other bit of jargon is the vomeronasal organ, or VNO. It is found in the roof of the mouth of many amphibians, reptiles, and mammals. It is nowhere to be found in fishes, birds, crocodiles, Old World primates, marine mammals, and some bats. Odors can enter this organ through the roof of the mouth and sometimes the nose. The forked tongue of the serpent evolved not to lie to Eve about the apple in Eden but to grab odor molecules from the air and then insert them into the VNO. A lot of pheromonal communication involves the VNO in animals that have it. Some researchers have argued that the seeming lack of a VNO in humans (we are not sure about this yet) indicates that we do not have pheromones. That logic is wrong because there is nothing that restricts an odor message to only the VNO—otherwise fish would not have pheromones, which is demonstrably not the case. We will not be concerned whether it is the VNO or the nose that helps to link external odors to our sexual brain.

Wherever olfactory receptors are found, they bind to different odor molecules that can be associated with quite different sources, such as rain, sex, and food. In mammals, these receptors are embedded in the mucous membrane inside our nose. Sniffing the air brings more molecules into the nose and increases our chances of smelling something in the environment. Mammals with a VNO expedite its odor-sensing capabilities with a flehmen response, as when a horse curls its top lip upward and bares its teeth. This is not exaggerated sniffing. The horse closes its nostrils when producing a flehmen response, which forces air into its mouth and channels the odors to the VNO. It is amazing that it is not known if humans have a functional VNO.

Olfactory receptors are usually neurons, and wherever they are found, in the toes or nose, they react to odors in the same way. When an odor binds to a receptor, it triggers a series of biochemical changes in the cell that eventually cause the neuron to fire. In mammals, these neural discharges are fed forward to the olfactory lobe of the brain. The olfactory lobe, in turn, projects to numerous other parts of the brain that are involved in different functions, sex being only one of them. Olfaction is alone among the senses in its immediate linkage to parts of the brain that are involved in memory, emotions, and the pleasures of "liking" and "wanting" governed by the mesolimbic reward system discussed in chapter 3. All of the other senses, including seeing and hearing, pass through relay stations lower in the brain for more processing before eventually projecting to pleasure centers. Odors waste no time and get right to the point. This direct linkage between olfaction and emotion is another attribute that makes this sense seem so primal.

Olfaction proceeds similarly in insects, and it is perhaps best understood in moths. Their olfactory receptors extend in netlike fashion from the antennae. There are two smells that are exciting to a male moth: the female's pheromone and the odors of the flowers it feeds on. The antennae contain an abundance of receptors that capture both of these classes of odors. When a receptor is stimulated by an odor, this stimulation is fed forward to a part of the moth's brain called the antennal lobe. Within the antennal lobe moths have groups of cells called *glomeruli*. Separate glomeruli act as a "sexual brain" or a "foraging brain" and are stimulated by different odors. The olfactory receptor cells that bind to sexual pheromones terminate in the macroglomerular complex, where the neural code for species and sex recognition is found. The olfactory receptor cells that detect floral odors, on the other hand, bypass this complex of cells and terminate in a region called the main antennal lobe; this is the moth's "foraging brain."

Sometimes sex and food are dependent on one another. Fruit flies often court on deposits of rotting fruit because this is where they lay their eggs. Olfactory receptors that detect food odors sensibly project to the flies' foraging brain. But one set of food-odor receptors where genes critical for male courtship (fruitless) are expressed go straight to the sexual brain.[1] If these receptors are not stimulated by food odors, then the male does not court. The interpretation is that male courtship will not ultimately be successful if the female has no place to lay her eggs. The male is not being chivalrous; he is being strategic.

Now that we know some of the details of this powerful sense, we can figure out how it figures into the sex life of so many and such different animals.

* * *

As I have pointed out many times, the most critical decisions about mating are doing it with correct species and with a partner willing and able. Olfactory cues are often the best for figuring that out. This is probably because odors are so closely tied to who we are and what we feel. The link between genes and odors is probably shorter and more direct than the same links in other signals.

Genes don't produce traits directly; there really are no genes "for" behavior. Instead, our DNA makes RNA, and our RNA makes proteins or regulates the action of other genes. This gets especially complicated when we ask naive questions about "genes for birdsong," for example. There are many aspects of physiology, morphology, and behavior that must be coordinated to result in a song. Genes must provide the blueprint for the construction of neurons that generate the specific rhythm of the song; genes have to direct the development of the cartilages, muscles, and bones along a very specific pathway that culminates in the bird's voice box; and genes must then somehow engineer the neural networks that connect all of these parts to get them to work in concert in order to produce those wonderful concerts diagnostic of so many songbirds. Coding genes to make a specific olfactory molecule is more straightforward. Genes orchestrate biochemical pathways that synthesize chains of compounds, and these compounds are the signals themselves. Pretty simple.

Not all odors come from genes, however. Environmental odors can act as olfactory fingerprints that reveal a good deal about ourselves. The odor of someone who has spent the night boozing and smoking in a bar provides an unmistakable fingerprint of this behavior. Other animals also use odors to track where individuals have been. Beehives are guarded by certain bees whose job is to admit only colony members.

If one takes a bee from the same hive as the guard bee, doses her with the odor from another hive, and then reintroduces her back to her own hive, the guard bee reacts as if the manipulated bee smells of smoke and booze, or worse, as if she is an alien from another hive.[2] The guard's reaction to the offending olfactory message, however, is a bit more extreme than ours would be to the stench of a barfly: the guard bee kills the messenger because of the olfactory message she now carries. When it comes to mate choice and these types of environmentally acquired odors, one main message is that if you smell like me, you must be like me. That is good when looking for the same species or, as we will soon see, not so good when looking for someone with different genes.

Odors can also reveal with whom we have been hanging out, and the consequences here can also be harsh. Many a relationship has been derailed by a man returning home to his partner bearing the scent of another woman. As I will discuss below, most women who wear perfume have a specific scent that they feel is "them," so when a philandering male returns home, the details of the alien scent does not matter, just the message that it is alien. But there might be details of a cheating devil in the scent's bouquet. Glendale.com, a dating site that arranges illicit affairs, reported the top ten perfumes used by cheating women and cheating men. If her man returns not only with a foreign perfume but specifically with a scent of Guerlain's Shalimar, Chanel's Coco Mademoiselle, or Givenchy's Very Irresistible, it might be time to get counseling.

Animals don't get counseling, they get right to the point. Female redbacked salamanders have low tolerance for a philandering mate. These small amphibians are common under rocks, logs, and mosses of forests in northeastern North America. They are ideal animals for behavior and ecology experiments, as they are perfectly content living in small terrariums. Experimenters at Southwestern Louisiana University documented the consequences of a male salamander going astray. Pairs of salamanders were maintained in separate terrariums mimicking what we might think of as marital bliss. But then the researchers interceded, planting doubts about the male's fidelity in the mind of his mate. They removed the male and placed him in another terrarium for a short time before returning him to his partner. If he had been placed in an empty terrarium, his return home was uneventful, but if had he visited a terrarium with another female, he had hell to pay. Even though the female was his size, she grabbed him mid-body with her jaws and body-slammed him onto the ground several times.[3] Lesson learned, although in this case it was the experimenter and not her mate who should have been the target of the female's wrath.

The environment appends some interesting phrases to the message of sexual scents by informing receivers of where an individual has been and with whom. Other odors more tightly influenced by genes provide different types of information. Fruit flies, fishes, snakes, and mammals all rely on olfactory signals that are coded in the genes to identify species, but moths are the champions when it comes to using scents for sex. In these animals, the typical roles of courters and choosers are reversed between the sexes; the females advertise their desires with odors, and the males key in on those odors to track down their mates. Females emit volatile pheromones that they waft into the air and that can travel with the wind for kilometers. When the male detects these pheromones, he follows the odor gradient upwind until he locates the solicitous female. Not only do the pheromones direct him; they focus him during his search for sex—too much perhaps. When the sexual odors stimulate his antennal lobes, he becomes oblivious to the sounds of bat predators who might be tracking him down.

There are many different species of moths broadcasting in the olfactory airwaves, similar to a surfeit of radio stations packing the available bandwidth of radio waves. Just as numerous, distinct FM and AM signals keep the radio stations separate on the dial, there is an abundance of odors that could be recruited to keep species identification error-free. Even though there are more than 100,000 possible different volatile pheromones available to moths, this would not provide a unique compound for each of the 160,000 species of moths worldwide. Moths do not, however, assign one unique odorant to each species. For example, 140 species of moths and the elephant all share the same primary sexual attractant in their recognition pheromones.[4] Confusion does not arise, however, because the different species of moths, like a growing number of wineries around the world, are into blends, changes in the ratios of different odorants provide another axis of variation for unique species identification. (Of course, there are other factors that keep a male moth from copulating with an elephant, more to do with smashing than blending.)

Most moths pick two odors for their sexual bouquet and emphasize one of them, yielding a blend with a major and minor olfactory component. Sometimes the major component for one species is the minor component for another. When moths evolve new signals, it is often by changing the ratio of the blend rather than adding or deleting a separate component. And these moths seem to have the latitude to evolve shifts in their blends as easily as a winemaker might shift the ratio of Malbec to Cabernet Sauvignon.

The cabbage looper is a well-known agricultural pest. As a caterpillar it wreaks havoc not only on cabbage but also broccoli, cauliflower, collards, kale, mustard, radish, rutabaga, turnip, and watercress. Given the number of folks who would like to see this moth dead, there has been extensive research into how it survives and reproduces. Its sexual pheromone is a two-part blend in a ratio of 100:1.[5] The moth has olfactory cells on the antennae that match each component and in the same 100:1 ratio. Thus, when the male whiffs his female's scent of 100 parts A to 1 part B, the neurons that code for A and B in his sexual brain, the macroglomerular complex just mentioned, fires in the same ratio. This is how their brains code mate recognition, and it is the site of their sexual aesthetics when it comes to the scent of sex. But what is attractive to one species is anathema to another. Because all 160,000 species of moths evolved from a common ancestor, and each species has both a unique sexual signal and a unique neural code to recognize it, there must have been a lot of evolution of both the signals and the receivers. How do these signals and the aesthetics for them evolve?

Evolution is often a slow, finicky process that we do not witness directly but rather deduce from patterns in nature: animals with thick fur live in cold climates, the length of a bat's tongue is precisely long enough to probe the flowers it pollinates, and bacteria are no longer threatened by antibiotics that have been in long and continual use. All of these relationships, we assume, are adaptations that result from evolution, even though we did not witness evolution taking place. Sometimes, though, we luck out, and evolution takes place right before our very own eyes.

In a laboratory strain of cabbage loopers, the evolution of a new species recognition signal and the neural code for it took place right under the noses of researchers. One day, lightning struck (metaphorically), and some mutant females started mixing their pheromone blends in equal proportions of the two olfactory components—50:50 (A:B) instead of 100:1. Initially, these females did not elicit much sexual attention from males, but then lightning struck again, and some males evolved a new preference, which made this mutant blend quite attractive.[6]

What changed in the males to make these normally oddly odiferous females smell sexy? The logical hypothesis is that the recognition code in the brains of the mutant males changed from 100:1 to 50:50. But as we have seen before, the logical solution is not always the biological solution, and this is yet another case. In these males, the code in the brain remained the same—neurons firing to A and B in a proportion of 100:1 signaled a desirable mate even though she was emitting her odors of A and B in equal proportion. How does this happen? In this case, the evolutionary action was at the receptors; the sensitivity of the receptors to component B changed, and the gain of these receptors was reduced by a hundredfold. Now it took one hundred units of odor B to elicit the same response from the olfactory-sexual brain that one unit previously elicited. The brain's code to recognize the mutant pheromone remained 100:1 even though the two components in the mutants were in equal proportion in the air and on the receptors. In both mutant and normal males, the same pattern of firing defined sexual beauty, despite being triggered by very different ratios of stimuli. Although the wild type and mutant females smelled quite different, there were equally beautiful to their own kind.

* * *

The opposite of speciation is hybridization. Although matings between different species usually do not produce viable offspring, it can happen; and when it does, it creates individuals that are neither the species of their mother nor their father but some mix. Hybridization can occur when we muck up an animal's sensory world. I mentioned fish biologist Gil Rosenthal as one my companions exploring the kelp forests off the coast of California. I introduced Gil to the natural wonders of swordtails in northeastern Mexico when he began his studies as my graduate student. Shortly thereafter, Gil introduced me to two wonderful species of swordtails in the mountains of Hidalgo, Mexico. On a stunning St. Patrick's Day, when the verdant hills of Hidalgo had a hint of Eire (although lacking the rainbow), a group of us hiked half a day over the mountains into a spectacular valley that served as the residence of two swordtail species, *Xiphophorus malinche* and *X. birchmanni*.[7] In this place, the two species obeyed the societal command reminiscent of the mournful refrain from the Janis Ian song "Society's Child," charging people to stick to their own kind.[8] They knew they were different species, and they acted that way.

But this wasn't true everywhere. Later, as a professor, Gil and his graduate student, Heidi Fisher, were working at a site on a larger river where the two swordtail species are found together. There as well, the two species had typically abided by the standard biological etiquette of no mating with heterospecifics. But one day Gil and Heidi found evidence that sexual bedlam had broken out—it was hybrids gone wild. *X. malinche* and *X. birchmanni* acted as if they didn't care, or at least couldn't recognize, with whom they mated.[9] The researchers surmised that this sexual indiscrimination might have something to do with an orange processing plant recently constructed upriver, whose discharges were polluting this site and causing eutrophication. They did experiments with these fish and showed that females could not discriminate between different and same species when tested in their home river water. When the same females were tested in clean water, however, they returned to typical biological norms and preferred males of their own species. Gil and Heidi realized that a by-product of eutrophication is humic acid, and that humic acid binds to olfactory receptors. Could the humic acid be blocking the ability of the female swordtails to discriminate mates? They tested females in clean water, and once again showed there was a preference for males of the same species, but when they added humic acid to the water, the female's discrimination ability disappeared. When the effect of the humic acid wore off, the females once again biased their sexual overtures to males of their own species. This all makes wonderful sense. We can't appreciate visual beauty in the dark, a melodious song masked by sounds of the city, or sexual odors if our noses are otherwise occupied. You can't desire what you can't sense.

Odors can inform choosers about a lot more than the species of the courter. As I am sure we are all tired of hearing by now, there is strong selection on choosers to mate with courters of the same species because their genes are complementary. The swordtails notwithstanding, making babies by combining genes from different species doesn't always work well. So a first priority in mate choice is getting a partner with similar genes. Within a species, though, not all genes are the same. I have blue eyes. If yours are brown, you have different genes for eye color (when we say different "genes" we usually mean different alleles or "gene variants" of the same gene). I am of Irish decent. If you are from the Middle East, we have some genetic differences, and we both differ from people in Asia, who have about 20 percent more genes from our Neanderthal relatives hanging around in their genomes.[10] But no genes vary like MHC genes.

The major histocompatibility complex (MHC) is a set of genes that function in our immune response. They identify cells from alien forms such as pathogens and parasites, and when they find them, they alert the body to recruit T cells to fight the invasion. The MHC genes need to be variable in order to accurately distinguish cellular friend from a huge diversity of foes, the "friend" being our own cells. This is why MHC genes are the most variable in all of the vertebrates. Because of all this variation, individuals can choose mates that yield offspring better armed to fight disease than either parent; that is, as long as choosers mate with courters who have MHC genes very different from their own. But how can we vertebrates tell if partners meet our MHC criterion?

Genomics allows us to have the MHC genes of prospective partners scanned and then compared to our own to determine our best match, at least in terms of MHC genes. I predict it will not be long before some overpriced dating services will require genome scans, and the first match they will use is MHC. But what is an animal, or a person without the resources to pay for a genome scan or more likely a genome *scam*, to do? Do we just decide that MHC is one of the inner attributes of a potential partner that we will only know when it is too late, like a volatile temper or a drinking problem? Do we have to wait for evidence that our child is immune-compromised to know we made a poor partner choice? No. It turns out that we are already paying very close attention to the MHC genes of potential partners; we just don't know it.

We can't see genes, but genes contribute to phenotypes. Eye color gives us a pretty accurate idea of the underlying genes that contribute to it. In other cases, the phenotype is not a very accurate predictor of genes, because many genes, rather than a single gene, might contribute to the phenotype, and the environment might play an overwhelming role in determining outward appearance. For example, genes influence body mass in humans, but so do beer, ice cream, and slothfulness. Looks can deceive, especially for the eugenicist.

There is, however, a phenotypic window that gives a good view of MHC genes—the animal's odors. This linkage between smell and genes is best understood in rodents in which odors in the "mouse urinary products" are correlated to MHC variation. Among species that have been studied, rodents with similar MHC genes smell similar to one another, while those with different MHC genes smell different. This sets the stage for a new criterion of beauty, not the MHC genes per se, but the scent they produce. This sexual aesthetic is a relative one. The fine details of a courter's MHC genes are not important, just whether they are different from the chooser's. MHC-based mate choice, however, occurs only in animals that use odor as an important criterion in mate choice.

What about us? We are somewhat sensitive to odors, and we all know how important scents are in courtship. We use flowers as a courtship gift not only because flowers look beautiful but because of their fragrant bouquet. There is a billion-dollar perfume industry, which we will come to later, that bottles fragrances to enhance our own bouquet. In addition, much of our behavior and physiology can be subconsciously influenced by odors. A classic study by Martha McClintock showed that the menstrual periods of undergrads in a college dorm tended to become synchronized over time.[11] The only logical cue was odor, and McClintock later identified the odor in question, the first human pheromone known to science.

We all know our behavior can be influenced by subtle cues of which we are consciously unaware. In a study relevant to the role of odors in sex, Geoffrey Miller, author of *The Mating Mind*, reported that men in strip clubs tipped more for a lap dance when the dancer was ovulating.[12] Although there were many uncontrolled variables in this study, such as the dancer's own behavior, Miller argued that the dancer's odors caused the enhanced generosity of their patrons. Although not experimentally verified, this conclusion does not seem so far-fetched, given McClintock's research into the correlation of odors and menstrual cycle. The "stinky tee shirt" experiment, however, is now the benchmark for showing the general link between smell and sex in humans, and specifically for demonstrating that MHC odors are an important criterion of our sexual aesthetics, one we don't even realize.

This is how the experiment conducted by Claus Wedekind and his colleagues works. Participants, male college undergraduates as is the case in many experiments on humans, volunteer to wear the same tee shirt for two consecutive nights. During this time, they refrain from bathing or using any fragrance such as perfumes, colognes, or deodorants. After this ordeal, they then put their tee shirt in a plastic bag and bring it to the laboratory, where women sniff the tee shirts and rate the attractiveness of the shirts' odors. In addition, the men and women have all previously been tested in order to determine their MHC type, and women report if they are using oral contraceptives.[13]

Women found the odors of men with MHC types different from their own odors more attractive than the odors of men with more similar MHC genes. Also as predicted from animal studies, the odor's attractiveness was not absolute—some guys did not generally smell better than others—but was context dependent; the odor's attractiveness depended on the woman's own MHC type. A clever experiment with dirty participants but clean results: women's percepts of beauty, like that of rodents, sticklebacks, and a myriad of other animals who sniff for sex, are influenced by MHC-based odors. (It is important to note, however, that these results have been replicated in some studies but not in others.)[14] The finding of MHC-based odor preferences in humans dovetails nicely with other studies that show a woman's sense of smell becomes heightened when she is sexually receptive, when mating is on her mind, and also that women rank a male's odor as the most important factor in choosing a sexual partner, whereas men rank the woman's looks as of paramount importance.

There is a slight catch to the stinky tee shirt experiment. The expected pattern of odor preferences applied only if a woman was not taking oral contraceptives. If she was on the pill, her preference was reversed; odors from men with more, not less, similar MHC genes were sexier. What is going on with the oral contraceptive? Why would a pill that influences cycles of reproductive hormones bias what smells good sexually?

Let's get back to the basic theory behind MHC-based mate choice. The prediction is that if a chooser can evaluate MHC genes, then she should prefer a mate with different MHC genes to produce healthier offspring. As noted above, it is MHC's role in immune function that makes MHC so variable. And this genetic variability is also a good indicator of how closely related we are to one another. Many animals use MHC variation as a clue to family heritage in contexts other than mating, usually when they are searching for assistance or wanting to share public goods. Frog tadpoles, for example, school with one another to reduce predation risk by means of a "selfish herd" effect; the more tadpoles hanging out together, the less likely you will be the one eaten by that hungry fish.[15] But they are not indiscriminate when sharing this benefit, as they prefer to herd with siblings over non-siblings. A good sniff of MHC-based odors lets them know if a neighbor is sib or non-sib.

Might females on the pill be more concerned about kinship than mating? The birth control pill functions by manipulating a woman's reproductive hormones to simulate pregnancy. Ovulation does not take place when a woman is pregnant, so women on the pill, if all is working well, do not get pregnant. Wedekind and his colleagues figured that a woman on the pill does not have mating on her mind, or at least does not have reproduction as a subconscious goal, so she is not interested in odor cues that indicate a good MHC-partner. Okay, this sounds fine, but then why do these women prefer the opposite, the odors of men with more similar MHC genes? Since the hormonal milieu of these women suggests that they are pregnant, then a post-pregnancy strategy kicks in. Identify those who can assist in child rearing, as we know it takes a village, in this case an unselfish herd, to raise a child. And who best to help in child rearing but close relatives? In some of today's mobile societies we are not always around relatives who are able to help out in family life. But the researchers correctly point out that the biology we have today, whether it is our morphology, behavior, or our sexual aesthetics, has a long evolutionary history and is sometimes better adapted to past conditions than present ones.

There is the potential for some unfortunate and unintended consequences to this interaction of mate-odor preferences and oral contraceptives, as pointed out by the researchers Fritz Vollrath and Manfred Milinski.[16] Let's say a couple is dating and the woman is on the pill. They fall in love, marry, and remain happy, and all this marital bliss leads to a decision to have a child. The woman stops taking contraceptives, and now the man she is sleeping with smells to her like her uncle! Not quite, perhaps, but now she is exposed to an odor of a man whose MHC is more rather than less similar to her own, and which she thus finds less attractive. We do not know if this scenario ever plays out in the real world, but it is a possibility that new couples might want to take into account.

* * *

The Darién National Forest, a UNESCO World Heritage site, is far from the crowded streets and bagpipers of that other World Heritage site I discussed, Edinburgh. Here, the raucous calls of red-and-green macaws replace the rhythmic pulsing of the bagpipes, and there are no red lights to tell you to stop or green ones to make you go. The Darién is a twelve thousand–square kilometer stretch of jungle in southern Panama along the border with Colombia. It is sometimes called "impenetrable," and this is one reason why it is the only gap in the forty-eight thousand–kilometer Interamerican Highway that stretches from Alaska to Argentina. It is also where Balboa landed on the Atlantic side of the Isthmus of Panama and marched to the other side to "discover" the Pacific Ocean.

The Darién Gap might be relatively impenetrable to some, but the indigenous Embera have been here since the late 1700s, when they displaced the native Guna to the San Blas islands and neighboring mainland. It is still burdensome to get around in the Gap; boats, horses, and hiking are the most efficient means of transportation. But there is nothing impenetrable about this forest when it comes to the chytrid fungus that killed off so many frogs in the mountains of western Panama and, we discovered, has recently arrived in the Darién.[17]

The Darién is one of the major hotspots for biodiversity in the Western Hemisphere, if not the world. One group of organisms that flourishes here is orchids. With long, simple, green leaves, they parasitize trees and usually are found high in the canopy. One way to see a lot of them is to find some trees that have fallen over and brought the canopy down to your level. This happens more often than you might guess. I have always been amazed by how many trees in the tropics just fall over, often in response to wind or a rainstorm. It is probably because the soil is so wet and thin here, which is also why so many trees have huge buttresses to keep them from toppling down. Fallen trees are critical in the ecology of the forests because they create gaps in the canopy that allow the usually dark surface to sparkle in the sun. The forest soil is a storehouse for a great diversity of seeds, defecated by birds and other animals that disperse them, and many seeds will not germinate until they have enough light. Pull back the curtains of the canopy to let some light in, and, presto, a variety of tree species begin to sprout. Light gaps are one of the most important facilitators of forest plant diversity.

Light gaps are also great places to find canopy species, be they arboreal frogs and insects, or parasitic orchids and bromeliads. We hiked through a light gap that opened up after a huge espavel tree crashed to the ground. This giant of the forest is in the Anacardiaceae family and can grow to fifty meters in height. When this one came down, it brought numerous smaller trees along with it, many of which were littered with orchids.

In chapter 3, I talked about how deceptive orchids exploit the sex drive of orchid bees to assist them in pollination.[18] The orchids accomplish this feat by evolving flower parts that resemble the silhouette and the fragrance of a female bee. In at least one case, the orchid's odor is even more attractive to a male bee than the aroma of a virgin female. In order to be deceptive, these plants have evolved to become top-of-the-line bee-perfumeries. Some bees, however, have turned the tables on the orchids by exploiting the plant's perfumeries for their own gain. They mix the fragrant odors from the orchid with a few drops of their own lipids to produce a greasy extraction similar to the enfleurage used in the perfume industry. The bees then suck up the perfumes from the plant and store them in body sacs for later use during courtship. In this bizarre web of nature, the male bees have engineered their phenotypes to make themselves more attractive to females by exploiting odors of orchids that have been engineered to be sexually attractive to bees so that the bees will help the orchids have more plant sex.

What happens if a male bee depends on orchid perfumeries for its courtship odors and then the orchids disappear? Such a situation has befallen some individuals of an orchid bee, *Euglossa viridisima*, which is native to Central America. Some of these bees were transported to Florida, where they thrived despite the fact that their orchid-perfumeries were nowhere to be found. But the bees didn't do without. Instead, they harvested odors from more than a dozen flowers that have components that allowed them to reconstitute the bouquet of their former orchid-perfumery. In fact, in their search to recreate their long-lost perfumes, the bees even become a bit creative, throwing in some basil to accent their aroma. So we see that humans are not the only species that rely on outside sources to enhance their sexual fragrance.

* * *

The greatest engineering feat in the history of human sexual beauty is perfume, and its role in human romance is legendary. Some aromas seem to go right to the sexual brain, where they immediately elicit liking and wanting. What determines the scents we engineer to enhance our own bouquets? Can we ask the engineers themselves?

Luca Turin, the subject of Chandler Burr's *The Emperor of Scent*, gives some entertaining and informative insights into the perfume industry.[19] Given how high the stakes are in this $5 billion business, one would think the theory of perfume would be well worked out. To the contrary, argues Turin. The industry is peopled by organic chemists who know which perfumes have been successful and take a rather random, hit-and-miss approach to assembling various compounds. These products are then tested by a panel of "noses," and most of them are deemed unacceptable. This method is costly and not very successful, as only a small percentage of bouquets ever make it to market.

Turin argues that if the perfume industry knew more about the biology of olfaction, it could use a bit of reverse engineering to increase its success rate. The problem, he argues, is that we don't know how olfaction works—or perhaps that he is one of the few folks who does know. He argues that it is not a lock-and-key mechanism where the structure of the odor "fits just right" into the receptor, but instead olfaction is based on the vibrational patterns of the molecules themselves whose detection is more akin to how we process sound. As of now, Turin's theory does not have strong support, but who knows, he might be right.

Regardless of how we sense perfumes, the theory had always been that we use them to mask unpleasant odors with pleasant ones. Not so, argues Manfred Milinski, who is the Director of the Max Planck Institute for Evolutionary Biology. Milinski did groundbreaking work on MHC and mate choice in stickleback fishes, so he has been contemplating the relationship between odors and sexual attraction for some time.[20] He has also focused his interest on humans.[21] He notes that some perfumes are reminiscent of some body odors, and then argues that the origin of perfumes is based on exploiting our olfactory bias to enhance our MHC-based odors. Given our ever-increasing understanding of the potency of MHC-based odors, Milinksi's idea about the origins of perfume makes sense. It seems logical, but is it biological, and how can one test this idea?

The most direct approach would be for us to characterize our own MHC odors and compare the chemical profiles of those odors to the chemical profiles of the perfumes we prefer. MHC odor research is not quite there yet, and, anyway, a list of organic compounds that comprise an odor does not necessarily tell us which of those odors are most salient. Milinski and Claus Wedekind, of stinky tee shirt fame, took a different approach.

In a study involving hundreds of men and women who were "fingerprinted" for their MHC types, individuals were given thirty-six different compounds that are commonly used to produce perfumes. They were asked to choose the fragrance that they would prefer to wear themselves, and the fragrance they would prefer that their partners wore. Even though we don't know what our MHC-linked odors smell like, we assume that folks with the same MHC genes have the same MHC odors. Thus the researchers predicted that folks with the same MHC genes should prefer the same perfume odors. And that they did.[22]

What about preferences for the odor of your partner, what would you like them to smell like? One prediction is that we do not want our partners to smell like us: it doesn't matter what your partner smells like as long as it is different. Remember, as far as we know, MHC-odor based preferences place a premium on genes and odors that are just different from ours, not different in any specific way. The preferences for these subjects in the odors they wished for their partners matched this prediction: they wanted their partners to smell different from themselves, but otherwise there was no consensus among folks with the same MHC genes as to what smelled good on their partners, despite the consistency as to what smelled good on themselves.

* * *

Plate 1. A calling male túngara frog at a typical breeding site in central Panama. The male's large vocal sac is as distinctive as his complex call.
Photo by Ryan Taylor

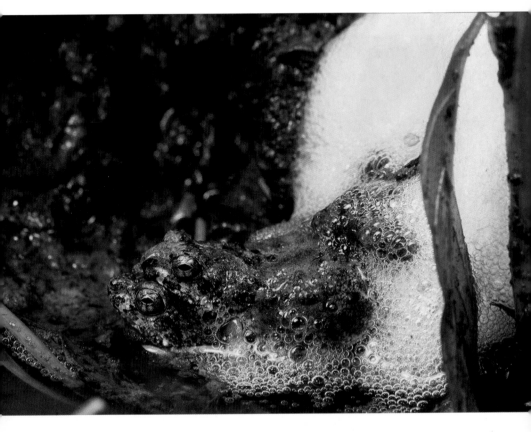

Plate 2. A nesting pair of túngara frogs. The male is on top as he clutches the female; they are in amplexus. The foam nest is produced when the male beats the jelly around the eggs with his hind feet. He is visibly exhausted after an arduous hour of nest construction.

Photo by Ryan Taylor

Plate 3. A fringe-lipped bat with a túngara frog. This bat can locate a calling frog by homing in on the call alone; the bat need not rely on its echolocation system to locate the frog. The túngara frog is its favorite prey in central Panama.

Photo by Merlin Tuttle

MerlinTuttle.org

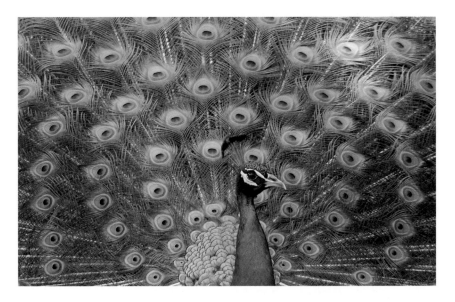

Plate 4. A male peacock erecting its tail feathers, or train, while
courting a female. Although the tail is enchanting to females,
it is this beautiful structure which, Darwin declared, made him
sick every time he saw it.

Photo by Jyshah Jysha

https://margotstaubin.wordpress.com/2014/10/20/pride-and-peacocks

Plate 5. The golden-headed lion tamarin is an endangered species found in lowland tropical forests in the state of Bahia, Brazil. It lives in social groups in which both males and females care for the young and for juveniles. Little else is known about its mating system. It is considered by some as the world's most beautiful primate.

Photo by Steve Wilson

Plate 6. Sexual selection often results in extreme differences between males and females. In many species the male is more adorned than the female, as seen here in the collared lizard: the more colorful male, top, is contrasted with the less colorful female, bottom.

Photos by A. K. Lappin

Plate 7. The peacock spider is a type of jumping spider; the male's colorful display is reminiscent of a peacock. His beautifully adorned abdomen is only raised when the male courts the female, at which time he waves it back and forth in an invitation to mate.

Photo by Jurgen Otto

Plate 8. The red bird of paradise is a native of Indonesia.
The males, one of which is shown here, are characterized by
a pair of long tail wires. During courtship these tail wires
seem to outline the male in the middle of a heart.
Photo by Tim Lehman

Plate 9. A male swordtail characin, right, a native of Trinidad, Taboga, Venezuela, and Colombia, extends its pectoral fin-ray with a piece of flesh that resembles a food item to the female, on the left. The female is attracted to this faux food item at which time the male initiates courtship.

Photo by Nicolas Kolm

Plate 10. An illustration of how a female greater
bowerbird might view a male bowerbird displaying
at his bower.

Illustration by John Endler

Plate 11. Male and female fireflies engage in spectacular nocturnal visual displays. As with many other courtship displays, the patterns of flashes are distinctive for each species. This image is a time-lapse photograph of synchronous fireflies from The Great Smoky Mountains National Park, near Elkmont, Tennessee.

Photo by Radim Schreiber

Plate 12. The quetzal is the national bird of Guatemala, and its image adorns the country's flag. Some consider the male resplendent quetzal to be the world's most beautiful bird. My binoculars began to shake in my hands the first time I saw one.
Photo by Dominic Sherony
Photographed at Savegre, Costa Rica, April 15, 2011, Resplendent Quetzal (Pharomachrus mocinno). Uploaded by Magnus Manske. Licensed under Creative Commons Attribution-Share Alike 2.0 Generic.

Plate 13. A male hairy caterpillar extruding his hair pencils. The tubes, or coremata, are inflated by blood pressure causing sex pheromones to be secreted through the hairs.

Photo by Rodney and Smudge Foster Rentz

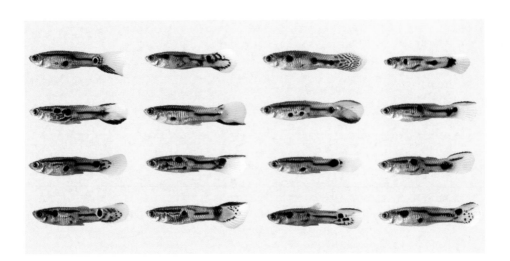

Plate 14. Guppies are known not only for their spectacular colors but also for incredible variation in those colors—as is the case especially of guppies in streams in Trinidad. Only a small sample of the striking variation is shown here.

Photo by Cara Gibson and Anne Houde

Plate 15. A bee orchid pseudo-copulating with an orchid. Although this behavior appears maladaptive, it makes perfect sense in the context of the bee's strategy for finding females. As females are few and far between, it behooves the male bee to copulate with anything resembling a female.

Photo by Nicolas J. Vereecken

Plate 16. A cricket, left, and the fly, *Ormia*, that parasitize this cricket. Just as frog-eating bats are attracted to the frog's mating call, these flies are attracted to the cricket's mating call. The fly deposits its larvae on the calling cricket and the larvae burrow inside, eating the male from the inside out as they develop.

Photo by Norman Lee

We have now explored the three major senses involved in our sexual aesthetics: seeing, hearing, and smelling. There are others senses used for sex by us and by other animals—tactile and electrical signals, for example. But it is these big three that are the best understood in terms of contributing to our sexual aesthetics and whose biology help us understand why we and other animals might perceive some as beautiful and some as not. But our sexual aesthetics do not act in a vacuum. In the next chapter, we will see how our social environment can have surprising and sometimes irrational effects on whom we find beautiful.

SEVEN

||||||||||||||||||||||||||||||||||||||||||||||||||||||||||||||||||||||||||||||||||||||||||||

Fickle Preferences

A woman is always a fickle, unstable thing.   —*Virgil*

In the previous chapters, we have seen how biases in the brain can influence our sexual aesthetics. Some of these biases have evolved because they steer the chooser toward a mate who is better: the right species, the opposite sex, disease-free, complementary genes, and more resources. In other cases, the biases for beauty exist for reasons outside of mate choice, and courters evolve traits that exploit these biases: appendages that look like food, calls that sound like predators, and courtship colors that stimulate eyes engineered to find prey. In all of these cases, we would expect biases for beauty to be stable—a peacock's tail should not appear diminished to a peahen only because it is Monday morning.

Fickle preferences, however, are common and might even be the rule rather than the exception. You can hear men whining Virgil's sentiment since time in memoriam. It is leveled as a criticism, but it merely means that when it comes to judging a man's attractiveness, women sometimes change their minds. But not only women and not only humans are fickle in their judgments of beauty, and there are many good reasons for being fickle. In this chapter, we will explore those reasons and ask why it is that percepts of beauty change in real time and not just evolutionary time.

* * *

Clocks keep ticking, and we know time flies. But we are usually oblivious to the power time has over how we perceive the world and how we make decisions about what we perceive. Our view of sexual beauty is especially susceptible to the tick-tock of time, and if we get an inkling that time is manipulating us, we usually deny it. When it comes to beauty we like to think we have our standards and that those standards are fairly stable. Our standards might change over the years but not so much over months or weeks or minutes. But change they do, sometimes in the wink of an eye.

There is no better repository of truths about the trials and tribulations of finding a sexual partner than country-western music. One of its stars, Mickey Gilley, provided a clever insight into our shifting percepts of sexual beauty in his song "Don't the Girls All Get Prettier at Closin' Time."[1] This song strikes a chord in many men because it reveals how fickle we are, how little control we have over it, and how we would rather deny it. Here is the story line: Mr. Gilley sings about a man in a bar longing for female companionship. When he checks out the prospects early in the evening, none of the patrons are quite up to his standards. This situation doesn't improve as closing time approaches, so now the lonely cowboy is faced with the reality of another lonely night. What can he do?

Mr. Lonely could solve his problem by lowering his standards; they are probably unrealistic anyway, which may be why he is so lonely. The lyrics suggest this is exactly what he does, but he does so at a price. When he awakes in the morning he is faced with the dissonance that comes from violating his sexual mores of settling for less: "If I could rate them on a scale from one to ten, / I'm looking for a nine, but eight could work right in. / A few more drinks and I might slip to five or even four, / But when tomorrow morning comes and I wake up with a number one / I swear I'll never do it anymore." Too bad for Mr. Lonely, because another solution would have been to maintain his standards of aiming for high numbers but to change his perceptions of beauty such that more individuals fall into categories eight and nine. In doing so, he could thus spare himself the guilt and embarrassment that comes with compromising his principles.

Gilley's song is entertaining. But it is more than just a song; the intuition it provides has motivated some real science. "Despite psychology's attempts at keeping pace with hypotheses generated by song writers, research dealing with perceived physical attraction has fallen far behind," Jamie Pennebaker and his colleagues wrote in 1979.[2] They remedied that situations with a study of how perceptions of beauty change as closing time approaches.

These researchers visited several bars in Virginia. Three times during the night they asked the patrons to score others of both the same and the opposite gender on an attractiveness scale of 1 to 10. The results were striking. The scores given by men and women patrons for members of their own gender tended to decrease a bit during the night, but their scores for members of the opposite gender shot up as closing time approached. This study, eponymously named for Gilley's song, corroborated his hypothesis that the girls *do* get prettier, or at least they *seem* prettier, at closing time—and so do the boys. One of Pennebaker's interpretations of these results was based on the psychological theory of dissonance, or as the authors put it, "If the subjects were committed to going home with a person of the opposite sex, it would be dissonant to consider an unattractive partner. The most efficient way of reducing such dissonance could be to increase the perceived attractiveness of the prospective alternatives." It is analogous to a professor curving the relatively poor grades on an exam and then becoming convinced he must be doing a great job teaching.

Pennebaker's study was repeated on the other side of the world in 2010 to address an uncontrolled variable in the original study—alcohol. A mantra of this book is that *beauty is in the eye of the beholder*, but we also know that sometimes *beauty is in the eye of the beer-holder*. The more recent study was conducted in Australia, where beer is king. The procedures were similar to the Virginia study, as were the results: perceived attractiveness of members of the opposite gender waxed as the night waned. But researchers in the Australian study also measured the blood alcohol concentration of those passing judgment on the beauty of their peers. There was a "beer-goggle" effect: folks looked more attractive when the scorers had more alcohol. However, this effect was still observed when alcohol concentration levels were controlled for, and the closing-time effect was still there pulling on our percepts of beauty.[3] Thus all of the efforts of nature and nurture that go into shaping our sexual aesthetics can be undone by the ticking of a barroom clock.

Not all clocks are in bars. All animals have their own biological clocks, and one that is especially hard to acknowledge is aging. The famed social-evolution theorist, Robert Trivers, in his book *Deceit and Self-Deception*, pokes fun at his own deceptions about his own sexual beauty, which are subject to this ticking. Trivers talks about chatting up an attractive young woman as they are walking down the street. He glances to the side and notices they are being followed by an old, decrepit fellow, graying, hunched over, and limping. Trivers quickens his pace, looks over his shoulder, and the stalker is still there. Only then does Trivers realize that he is the stalker—he had been seeing his own reflection in the store windows.[4] The presence of a young attractive woman made him think and feel that he was younger, so much so that for a moment he was unrecognizable to himself.

The ticking of the biological clock in women receives a lot more attention than it does in men. Women have two clocks running simultaneously that are tied to reproduction and, not surprisingly, influence how they engineer their beauty and modulate their desire for sex. The first clock governs the reproductive cycle. In chapter 5, we discussed how this cycling of reproductive hormones in White-crowned Sparrows influences both liking and wanting sex. The same thing happens in women about every month. As with all vertebrates, the time during the reproductive cycle when eggs are ovulated and available to be fertilized is limited. In previous chapters, I gave examples of how humans and other animals engineer their beauty. If appearing attractive helps to get mates, and the function of mating is to get your eggs fertilized, then one might predict that women should be more attentive to their physical appearance during ovulation. The evolutionary psychologist Martie Haselton and her colleagues made just this prediction.

Their approach to testing their "fertility-ornamentation" hypothesis was simple. They photographed women when the women were in both the fertile period and the nonfertile period of their menstrual cycle. The researchers then showed these photographs to judges who were asked to rate which of the pair of photos of the same woman suggested she was trying to appear more attractive. There was a significant effect in the expected direction. In general, women in their fertile period appeared "more fashionable, nicer and showed more skin" than did the same women when in their nonfertile period.[5] Fertility ornamentation is not restricted to visual cues; in another study Haselton showed that women also spoke in higher-pitched, more feminine voices during the fertile period.[6] Finally, it is not only what women do to themselves but what they do to other women when they themselves are fertile: they are more critical of other women's attractiveness and less likely to share monetary rewards with them. As Haselton and her team pointed out, all of these results might explain previous studies that show men are more possessive of their partners in the fertile period (in animals, we call this *mate-guarding*). Alternatively, it might just be that this is a time when men have more to lose, whether or not women are advertising their fertility.

The second biological clock is that of aging, and it ticks down relentlessly to the final closing time for all of us. In terms of reproductive potential, it ticks with more urgency in women as they approach menopause. A man's sperm can remain viable, even if his spirit is less willing, throughout much of his life, although we now know that genetic mutations in the sperm increase with age and the male's ability to fertilize eggs decreases with his age.[7] Once a woman is in her twenties, however, her fertility decreases with age until she reaches menopause, at which time reproduction is no longer an option. But women don't take this pressure lying down. According to Judith Easton and her colleagues, "women have evolved a *reproduction expediting psychological adaptation* designed to capitalize on their remaining fertility."[8] What might this fancy-named adaptation be? It is a pretty simple one: middle-aged women fantasize more about sex and actually have more sex than their younger cohorts. The interpretation is that when time is running out, be it in a bar or in one's reproductive life, it is no time to be too fussy.

Let's not close this discussion of closing times with the idea that it is only humans who watch the clock. Kathleen Lynch investigated this phenomenon in túngara frogs and showed they too have standards of beauty that change with time. As we found out when we visited one of their orgies in chapter 2, a female túngara frog only shows up at the sexual marketplace on the night she is ready to mate. If she doesn't mate that night, all the eggs she ovulated are wasted; those genes never make it to the gene pool but instead are flushed out of the female's reproductive tract to become food for the fishes and insects that swim in her breeding pool. Does the female túngara frog also have a "reproduction expediting psychological adaptation" to ward off such waste? She does, and one would think she also had been taking her cue from Mickey Gilley. Lynch tempted females with a synthetic mating call that deviated substantially from the normal call of a male túngara frog, and which other studies had shown to be unattractive to them. Early in the evening, females were attracted to the normal call when it was broadcast in the testing arena, but they usually ignored the deviant call. Later in the night, when females were approaching their own closing time, their standards of acceptable beauty had changed, and now they were quite accepting of this deviant and usually unattractive call; they even responded to this call faster than they would have to an normal call earlier in the evening.[9]

Female túngara frogs are not alone in this type of fickleness; other female animals respond similarly with age: older roaches need less courtship before they decide to mate, and guppies and house crickets become less choosey with age. All these animals become more permissive as closing time closes in, perhaps reducing their "dissonance" along the way.

In humans, both sexes can vary their sexual strategies with creeping age: they can enhance their sexual beauty with ornaments or just kid themselves about how attractive they are. In animals, there are fewer examples of males responding to the steady approach of the grim reaper. An illuminating example comes from the males of the common fruit fly, *Drosophila melanogaster*, for whom death comes too quickly—thirty days, and they are gone.

Male fruit flies have mature sperm a mere two days after they emerge from their pupal case, but they are less fertile than their elders of seven days. These youngsters are also at a disadvantage when they compete with their elders for females. For both male and female fruit flies, mating makes babies, but it also shortens their lives, owing to, it is thought, all of the energetic costs involved in courtship and reproduction.[10] So there might be some advantage for young males to forsake mating until they are older, when sex will more likely lead to offspring. There are numerous ways to avoid sex when young, but as we know from our own species, abstinence is not an especially effective one, especially when the will is weak. The evolutionary solution in fruit flies is that younger males don't notice females as readily, compared with the more sensitive older males.

In the previous chapter, we talked about the olfactory receptor neurons (ORN) in moths that are involved in courtship. These are the ORNs that express the gene *fruitless*, a gene that plays a key role in courtship. *Fruitless* is also in fruit flies—in fact, it was discovered in these animals. ORNs that express *fruitless* are also key to courtship in fruit flies, where they go by the sexy name of OR47b.[11] When researchers arrange sexual competitions between seven-day-old and two-day-old males, the older ones have a 2:1 mating advantage.[12] Could this be because more mature males can better detect females? Researchers addressed this question by knocking out the genes that are responsible for the OR47b neurons. They then placed the mutant seven-day-old flies in competition against normal flies of the same age: the normal males got more matings. Thus OR47b neurons are important for older males to get matings. Is the same true for younger males? When the same experiments were repeated with the younger males, the results were different: two-day-old mutants with no OR47b receptors did just as well as the normal two-day-old males. Even though two-day-old males have these neurons, they don't seem to make any difference in mating success.

These results suggest that OR47b neurons account for the difference in mating success between younger and older males. Why? An obvious guess is that these neurons are more mature and more sensitive in older males. To test this idea, the researchers conducted neural recordings of OR47b, very much as we did with túngara frogs as described in chapter 2. Instead of playing sounds for an animal, they blew female odors across the males' receptors and found the OR47b receptors to be more than one hundred times more sensitive in seven-day-old than in two-day-old flies. In this case, the response of old age to sexual opportunity is motivated by different concerns from most of the examples just discussed. The male fruit flies have evolved increased sensitivity to females when they are older in order to curb their sexual appetite when they are younger. When younger, a male is less likely to be successful in attracting a female, and this effort might actually increase his mortality risk.

In this example, at least, older males are not wiser, just more sensitive. Internal clocks can explain a lot of the variation we see in what animals, including us, find sexually attractive. So the next time a love interest appears fickle, just check the time.

* * *

It is not only our internal biology that makes us fickle; there are external forces at work as well. We would like to think that we are originals, unique members of our species. Technically, this is true; no two individuals are exactly alike. But much of what is "us" is copied from others. Our genes are copies of our parents' genes; our language is copied from others when we are young; and our tastes in music, art, food, and the sports teams we root for are cultural norms copied from those around us. Furthermore, teenage sex, marijuana and alcohol abuse, and a realm of irritating adolescent behaviors are all partly blamed on the inability to resist copying peer behavior. Much of this makes sense. When animals are social, and we are among the most social, there is plenty of public information out there, and sometimes it is to our advantage to attend to it. If someone is successful, let's do what he or she did and hope for the same outcome. If we want to fit into a group, there is always pressure to be like other group members. Not all peers are created equal, however, and the pressure of peers varies, depending on who they are. We all know that careful choice of our associates is one way to extend and enhance our phenotypes.

A wonderful example of this is played out in the movie *Legally Blonde*. A snooty young woman blows off the request for a date from an awkward, nerdy young man—"women like me don't date losers like you." The charming, beautiful, down-to-earth Elle Woods (played by Reese Witherspoon) happens to hear the conversation and feels for the poor guy. So, she approaches him and in tears asks this stranger how he could have broken her heart like that. She then marches off faking a broken heart. After Elle leaves, Miss Snooty, who has been eavesdropping, goes back to the guy and asks, "So when did you want to go out?"

The notion that our view of beauty is influenced by the sexual aesthetics of others is called *mate choice copying*. Although it is hardly a surprise that this occurs in our own species, mate choice copying only became a serious topic for evolutionary biologists when researchers were trying to figure out why so few males garnered so many of the matings on sage grouse leks. Leks are the most extreme sexual marketplace in the animal kingdom and characterize the mating system of a diverse array of animals. Males gather in a specific area, the lek, for the sole purpose of advertising their sexual wares to females, and females are in charge in deciding which males get to mate. Paradoxically, only a few males are successful (and very successful at that), but the difference between the males in terms of things researchers can measure, such as their size, age, plumage colors, and mating displays, are not all that different. How can small, or even nonexistent, differences in the males' appearance lead to such large differences in their mating success?

Sage grouse leks are peculiar places. These birds live in the sagebrush country of North America, and there is nothing outstanding about where their leks occur. But although they are seemingly located at random, the leks occur in the same place from year to year. Records from Native Americans show that some of these sites have been in use for more than one hundred years. My first encounter with a sage grouse lek was in Wyoming early one morning while the stars still filled the sky and the temperature skirted freezing. As the sun began to rise, I was treated to the sight of dozens of males strutting around with their pin-like tails erect and their chests puffed out; two yellow sacs rocketed out from their white chest-feathers as the males produced odd swishing sounds. A few females moved through the lek in no apparent hurry as they surveyed the courters. Female sage grouse have complete freedom to choose their own mates, but it seems they are not all that confident in making their own choice.

A female grouse gets nothing more from her mate than sperm; he will never be a parent, good or bad, and he offers neither food nor protection to his mate. Although a large number of males appeared to the researchers to be equally attractive, only a few were chosen as mates by females. Consequently, each year fewer than 10 percent of the males completed more than 75 percent of all copulations, but there didn't seem to be anything about these males that could account for their extreme attractiveness.[13] So how is it that a few male sage grouse, not that different from the rest, are viewed by most of the females as being the most attractive?

The paradox disappears if we entertain the possibility that female grouse don't always think for themselves. What if mating decisions are not made independently by females, but are influenced by what others find attractive? Consider this scenario: there is a group of equally good-looking males, but when one of these males is chosen by a female, he becomes even more attractive in the eyes of onlooking females, who then copy the choice of their peers. When this happens, the chosen male is off to the races; the more he mates, the more attractive he becomes, which delivers him more mates, making him even more attractive to the copycats. Mate choice copying seems to be a logical solution to this paradox, but is it the biological solution—that is, does it really happen? Although studies of sage grouse motivated the explanation of mate choice copying, these birds are not ideal subjects for experimentally testing its validity. But fish are.

Guppies are one of the most variably patterned vertebrates. They sport a variety of colors splattered over their bodies. But orange reigns supreme in the eyes of the females. Much like the eyes of the surf perch discussed in chapter 4, the heightened sensitivity of guppies to certain colors evolved in response to the food they must see; in this case orange fruits that fall into the water. Male guppies, it has been argued, evolved orange ornamentation to exploit this sensory bias in females. But it is not only the colors she sees in males that influence the female's choice, it is also what she sees other females choosing.

The biologist Lee Dugatkin used a simple experiment to show that female guppies show mate choice copying, as was suggested for sage grouse and portrayed by Reese Witherspoon. Dugatkin placed a female guppy in an aquarium with two males, each in a separate compartment on opposite ends of the female's home tank. The female could go to both sides to court each of the males. The amount of time she spent courting each male indicated her relative attraction to him, and as others had shown before, she usually preferred the male with more orange. Dugatkin then placed the test female back in a transparent container in the middle of the tank, and placed a female, the "model," in the compartment of the less-preferred of the two males. From her container, the test female could watch this male court the model female. The model was then removed, and the preference of the test female, who had just been a voyeur of the less-preferred male's sexual activities, was retested. Her initial preference proved fickle; she reversed her preference and now spent more time with the male who had been shunned earlier.[14] These results offered one possible resolution to extreme skew in mating success in other species—even if all males are equally attractive, one has to be chosen first, and if females are copycats, the chosen male will enjoy a substantial proportion of all the matings. There are now numerous studies that have demonstrated the potency of social context in influencing a female's percepts of sexual beauty.

Mate choice copying is not restricted to guppies, and understanding this helped some of us resolve another paradox in another fish. Sailfin mollies are typical fish in that males and females reproduce by mating with each other. A similar-looking species of fish, the Amazon molly, however, consists entirely of females. It is named for the all-female tribe in Greek mythology whose only contact with men was for procreation. Amazon mollies are somewhat similar in that they also still need males. Although Amazon mollies produce clones of themselves with unfertilized eggs, they still need sperm to get their eggs to develop; the sperm does not fertilize the egg but delivers some sort of a biochemical nudge to get the eggs to start developing. Given this burden of needing to mate with a male, Amazon females are in a bit of a quandary, since there are no male Amazons. So what to do? The female's solution is to find a male that would be most like a male Amazon if Amazon males were to exist.

Amazon mollies resulted from an evolutionary mistake. They evolved about three hundred thousand years ago in Tampico, on the Gulf Coast of northern Mexico. This mistake, which gave rise to a brand-new species, occurred when a female shortfin molly mistakenly mated with a male sailfin molly. In rivers north of Tampico in Mexico and in Texas, Amazons live together with sailfin mollies; south of Tampico, they cohabit with shortfin mollies. Depending on where they live, Amazon females use either male sailfin or male shortfin mollies as sources of the sperm to get their eggs on the road to reproduction.

Scientists had understood this bizarre mating system from the Amazon's perspective for some time, but I became interested in it because I was troubled by not understanding why male sailfins would agree to be part of such an odd couple. I was troubled not because I have anything against matings between different species—they usually don't work out, but if fishes want to try, maybe they will be the exception—but because

I did not understand what the males got out of this. There is always a cost to mating: energy invested, time wasted, and predators attracted. For males, who usually try to mate as much as possible, the costs of mating are usually far outweighed by the benefits of successful fertilization. But there is no chance of making babies who carry any of the male's genes when a sailfin mates with an Amazon. The endeavor seems a total waste. Nevertheless, remember the orchid bee in chapter 3 who helps plants have sex? I wondered if, on closer examination, male sailfins were behaving in a manner that made adaptive sense. Maybe there was some subtle, hidden advantage that these males extracted from their dalliances with the Amazons.

Scientists had given only fleeting attention to the seemingly maladaptive behavior of these males; the consensus was that they were stupid and horny. The fish either could not tell the difference between their own sailfin females and Amazons (stupid), or they just didn't care (horny). I was sure about the horny part—they are males, but I doubted that they were really that stupid. If I could tell apart female sailfins and Amazons, certainly the sailfin males could. We "asked" the males if they could discriminate by placing a male in a tank with a sailfin female and an Amazon and counting how many times the male tried to insert his intromittent organ, the fish's equivalent of a penis, into each female. Although sailfin males would mate with both types of females, they showed a strong preference for their own: horny but not stupid. So why bother with Amazons?

My postdocs Ingo Schlupp and Cathy Marler and I wondered if the recent studies by Dugatkin might hold the key to this odd pairing of sailfins and Amazons. Might male sailfins benefit from mating with the Amazons because of mate choice copying making them appear sexier to their own females? Our experiment was similar to that of Dugatkin's. A female sailfin was given a choice between two sailfin males. Inevitably, the female preferred one male to the other, usually the slightly larger one. We then let the test female witness the less-preferred male courting a female Amazon. Would a female sailfin copy the mate choice of an Amazon female? She did. When her preference was again tested, the female sailfin now found the previously less-preferred male more attractive. Mate choice copying took place even though the sailfin female was copying the mate choice of another species.[15] Male sailfins might be wasting their sperm when they mate with Amazons, but they are not wasting their time; they are enhancing their attractiveness.

The female sailfins also have an added twist to their copying behavior, one that was revealed by an evolutionary psychology student at my university, Sarah Hill. Sarah was interested in human mating behavior and would often come to our weekly lab meetings where we talked about animal sex. She was a bit frustrated that she could not do the types of experiments with humans that we can routinely do with fish. So she added fish to her repertoire of studies.

Sarah wanted to know if the quality of the model influenced the degree of mate choice copying by a female. Again, the sailfins and Amazons offered a good study system. As mentioned above, males will mate with Amazon mollies, but they prefer sailfin mollies; in their eyes, their own females have higher "utility" than Amazons. If a female is confronted with two males, one courting an Amazon female and the other courting a sailfin female, we—and more important, the voyeur sailfin female—can assume that the male with his own female is more attractive than the male with the Amazon female.

Sarah repeated what had by now become the standard mate choice copying test. In her experiments, however, after the female sailfin chose a male, both of the males were given models. The preferred male was given an Amazon female while the un-preferred male was given a sailfin female. Thus the utility of the models varied. We predicted that the sailfin model should skew the female preference to the un-preferred male even though the preferred male was also consorting with a model, just a less desirable one. And this is exactly what happened.[16] It is not just having a consort that influences your attractiveness, but also how attractive that consort is.

Most of us probably do not need to be convinced that mate choice copying occurs in humans, and there are plenty of data to support our intuitions. Most of these psychology experiments follow similar procedures, but the participants are WEIRD, so let me digress. A constraint on researchers studying humans, one which confronted Sarah Hill, is that there are many types of experiments that they are ethically barred from performing. Overall, this is a good thing, but it does cramp the scientific style of psychologists who study humans. An alternative approach to an experiment is the questionnaire, but asking questions only works if you have someone to answer them. Luckily for many psychologists in academia, they have a captive audience—undergraduates who enroll in classes in which they are required to participate in studies as part of a class or for additional course credits. As the psychologist Joe Henrich and his colleagues pointed out in an article published in *Behavioral and Brain Sciences*, the participants in most of these studies are WEIRD (Western Educated from Industrialized Rich Democracies).[17] Well, okay, they might be weird, but they are still human, so can't we extrapolate findings based on them to other humans? Yes, to some extent, but we must remember that these subjects are usually from countries that in composite represent only 12 percent of the world's population. Also, for the most part, these subjects are teenagers or in their early twenties. This means their brains are not yet fully developed: they tend to be more risk-insensitive, more interested in immediate gratification, and somewhat limited in their life experiences. Finally, there is reason to believe that the subjects are not always deeply committed to giving true, unbiased answers to the questions they are asked. But they are what they are; we just need to remind ourselves that they are not everyone and that results from WEIRD subjects might not apply across countries, cultures, classes, and ages. For example, in chapter 4, I briefly mentioned that a woman's waist-to-hip ratio, independent of body mass, influences her attractiveness to men: 0.71 is the ideal. Most of the studies that contribute to this result are WEIRD-based, and nearly all of them were conducted with subjects who have been exposed to Western culture through mass media. The anthropologist Lawrence Sugiyama measured men's preferences for women's waist-to-hip ratio and body mass in a remote tribe of the Shiwiar in Amazonian Ecuador. Here he discovered that men found body mass more important than waist-to-hip ratio.[18] These results do not negate other studies showing the importance of waist-to-hip ratio, but they do show that there probably is an important cultural influence on this particular aspect of female beauty. As cultures vary, so might their sexual aesthetics.

Studies of humans revealed the importance of mate choice copying even before evolutionary biologists gave it that name. The psychologists Harold Sigall and David Landy anticipated much of the research on mate choice copying in 1973. They asked, "Do the impressions created in bystanders contribute to our desire to have relationships with beautiful others?" They conducted experiments in which a subject entered a waiting room containing two other undergraduates, one man and one woman. The man, the target, was of average attractiveness, and he was paired with a woman, the model, who was either attractive or unattractive, each condition being further enhanced by the model's clothing. The subjects were later asked to rate their overall impressions of the man and the degree to which they might "like" or "dislike" him. Men were rated more favorably when they were paired with an attractive woman than with an unattractive woman.[19] Although these studies were not explicitly evaluating sexual attractiveness, the hint is there.

More recent studies in the specific context of mate choice copying have shown similar effects. David Waynforth, an anthropologist, showed undergraduates photographs of men by themselves and then with a date, either an attractive or unattractive one. The subjects were asked to rate the attractiveness of the man's face. A male's attractiveness increased if he was with an attractive date.[20] Sarah Hill, returning to humans after her short foray into animal sexual selection, showed that the utility of the model in human mate choice copying is not categorical—that is, attractive models promote while unattractive models detract from the target's attractiveness—but that the effect scales more continuously as the model's attractiveness varies.[21] It seems we do not pigeon-hole others as either attractive or unattractive but judge them on a spectrum of more or less attractive. However, we are judgmental!

Just because a man becomes more attractive in the eyes of a voyeur when he is with an attractive partner does not mean he is seeking out this effect. But he well might be. Trophies usually document the winners of a competition. Although we often speak of competing for a trophy—for example, "our team will bring home the trophy!"—the trophy is only being used as a metaphor for winning the competition; the trophy documents victory. As Jarod Kintz writes in *This Book Is Not for Sale*, "A trophy isn't about the hardware, the gold-painted statue mounted on marble, it's about the recognition of excellence. A trophy is a physical representation of the abstract concepts of hard work and dedication. And that's precisely why I don't have any trophies."[22] It is in this sense, I think, that some use the derogatory term *trophy wife* for a much younger, attractive spouse of an older, usually rich, man. She documents his success in the competition of life, or at least the competition of making money, and men love to flaunt their trophies!

There is more than hearsay evidence that men are conscious of the "utility of their model." Do they knowingly flaunt their attractive partners? In the study just mentioned by Sigall and Landy, the male targets reported that they thought they would be rated more favorably when they were with an attractive model than an unattractive one, especially if an onlooker thought that the attractive model was a girlfriend. A more recent study shows that not only do men know this, they flaunt it. Undergraduates in Missouri were told they would be handing out pamphlets on campus while paired with a partner of the opposite sex with whom they were to pretend to have a relationship. The subjects were given a photo of their prospective, but fictional, partner and were then asked to decide if they wanted to hand out these pamphlets in areas frequented by undergraduates or areas with administrators. The "flaunting" hypothesis predicted that men matched with attractive partners would want to "flaunt" their partners to their peers, but if they were assigned unattractive partners, they would want to "conceal" these partners by working the area with administrators rather than other students. Both men and women followed the axiom "if you've got it, flaunt it."[23]

We need to be a bit cautious in interpreting some of these studies on human preferences for beauty. First, as I mentioned, the studies tend to use subjects from one particular demographic that need not be representative of humans in general. Second, the studies often use proxies that are somewhat removed from the general phenomenon being investigated. For example, preferring the photograph of one individual over that of another individual need not be indicative of how individuals choose mates. Third, just because humans and fish, as well as some other animals, exhibit mate choice copying does not necessarily mean that this feature evolved under the same selection forces, that it is influenced by the same mix of nature and nurture, or that it serves similar functions in different species. Nevertheless, studies of evolutionary psychology are asking important questions about why we are who we are.

In this section, we have seen that our social peers influence how we are perceived: their attractiveness creates a halo effect on our own attractiveness. This seems logical, although it is a fairly recent finding that sexual aesthetics can be so socially pliable. In the next section, we delve into a more recent, and seemingly illogical, effect of social context on how we perceive beauty.

* * *

Crazy is a word we often hear associated with being in love, being in lust, and falling for someone. Consider the titles of some songs: "Crazy Love," Frank Sinatra; "Crazy in Love," Kenny Rogers; "Crazy for This Girl," Evan and Jaron; "Crazy Stupid Love," Cheryl Cole; "Crazy Wild Desire," Webb Pierce; "I'm Crazy 'bout My Baby," Marvin Gaye; and "Crazy in Love," Beyoncé; it sounds like a playlist for an asylum! Crazy and love might go together because we often do not understand just what someone finds attractive in another person.

The opposite of crazy is sane. When people are sane, we assume they are rational, but when people are irrational we don't assume they are insane. This is a good thing, because irrationality is rampant, and we hope the inmates are not running the asylum, in our species or in any others. Yet we know little about how rational our percepts of sexual beauty might be. To explore this turf, I need to explain what I mean by *rational*, and here I draw on the insights of economists rather than philosophers, as the former provide more quantitative substance. According to classic economics dogma, individuals behave rationally when they behave in a manner that maximizes some utility. Economists assume that we strive to maximize economic gain; evolutionary biologists, who have bought heavily into classic economic analysis, assume that animals strive to maximize their Darwinian fitness.

How do we know if an individual is behaving rationally, since usually we cannot predict the amount of money or the number of meetings they will acquire in the long term? Two important criteria are that a rational individual is one who makes choices that obey the simple mathematical axioms of transitivity and regularity. Transitivity assumes that if A > B and B > C, then A > C. Transitivity is common in our world. If Lucy is taller than Emma, and Emma is taller than Gwen, we don't need a ruler to know that Lucy is taller than Gwen. Transitivity is a useful rule that increases our information about relationships. But transitivity is often violated. The children's game rock-paper-scissors is an example of intransitivity: rock beats a scissors by crushing it; rock loses to paper because it gets covered by paper; but paper's dominance is short-lived because it is beat by scissors: snip, snip, snip. Intransitivity is common in adult games as well. Consider a common gambler's fallacy: team X beat team Y a few weeks back; last week team Y beat team Z; therefore, X should beat Z this Sunday (X > Y, Y > Z, so X > Z). X is sure to win, right? You can put your money on it. The fallacy that sports are transitive has made many a bookie rich. As they say in American professional football, this is why they play the games on Sunday.

Many theories about how mate choice evolves assume that it is transitive. As far as we know, and we don't know all that much, that faith is well placed when it comes to our perceptions of sexual beauty; transitivity has been demonstrated in zebra finch preference for bill color, pigeon preference for plumage pattern, and cichlid preference for body size. A slight exception is a study of transitivity in túngara frogs, in which Stan Rand and I conducted the experiments and our colleague Mark Kirkpatrick analyzed the data with some of his mathematical wizardry. Females were given choices between all possible pairs of nine male mating calls. The results of that study showed that the female frogs tended to be intransitive.[24]

There is a surprising dearth of studies of transitivity in human mating preferences. But when addressed, transitivity seems to be supported. A study by the evolutionary biologist Alexandre Courtiol and his colleagues, for example, showed partner preferences for height in humans: women preferred men taller than themselves, and men preferred women shorter than themselves. There was a ceiling and floor effect for both genders; people became less attractive as partners when their height approached the ceiling or the floor. But within the considerable range in which height mattered, preferences for it were transitive.[25]

The other criterion of economic rationality is regularity. This occurs if the perceived relative value of A versus B is not influenced by the addition of an inferior third party, C. I exhibit regularity when my choice between a Czech lager and an IPA is not influenced by whether the tavern also serves Coors Light—and, believe me, it is not. But regularity is not always the case; not only is regularity often violated, it is used against us in the world of commerce. A well-known violation is the asymmetrically dominated decoy, or competitive decoy, effect. Here is an example, and consumers beware.

You are shopping for a car, and you are interested in a vehicle that has both a low price tag and high fuel efficiency. Your car dealer shows you two models: car A gets twenty-five miles per gallon (mpg) at a price tag of $25,000, while car B gets only fifteen mpg but costs less at $20,000. How to decide? As you ponder the relative costs and benefits, the salesperson decides she will make the choice for you, or more precisely she will manipulate you to think you made the choice for yourself. Watch out . . . here comes the decoy. The salesperson is bubbling over with statistics and good-natured banter as she shows you a third option, car C: twenty-two mpg at a whopping $40,000. Car C is way out of your price range, and she knows it. So how is she manipulating you by offering you a product she knows you will not want? All of a sudden your decision becomes easier. You now decide to purchase A. You get the best fuel efficiency of the three at a good price—at least a good price compared with the decoy, although you now choose the more expensive car of the two you originally considered. And by the way, the salesperson's commission is based on how much the car costs, not how far it can drive on a gallon of gas.

So how does this work? One interpretation is based on the name of the phenomenon: *asymmetrically dominated decoy*. The decoy, C, is a poor choice compared with both A and B, but A is superior, or dominates, C on both value metrics, gas mileage and price, while B dominates C on only one metric, price. Thus A is the superior choice. Another explanation has a more perceptual basis. When comparing the costs of the vehicles, B was better than A: $20,000 compared with $25,000. But when C was introduced, it expanded the range of prices being compared—$20,000 to $40,000—thus the difference between A and B, a mere $5,000, no longer seemed that substantial compared with the $15,000 difference between A and C. Since the gas mileage of C, twenty-two mpg, was between that of A and B, the range over which gas mileage was compared stayed the same, from fifteen to twenty-five mpg, as did the perceived difference between A over B. Again, A is the superior choice but for a different reason.

Given how competitive decoys so easily influence human behavior in the economic marketplace, it is not surprising that that their effects extend to the sexual marketplace. The social psychologist Constantine Sedikides and her associates, including Dan Ariely, the author of *Predictably Irrational*, queried students about what they found attractive in partners. In one of their experiments, subjects reported their preferences among three male models; the targets were neither real nor photos, but rather descriptions of the attributes of each male. Males A and B were always two of the three in the triad: A was physically more attractive than B, and B had a better sense of humor than A. Subjects were given one of two triads to evaluate, each with A and B, and a competitive decoy, C. There were two variants of the competitive decoy. In one triad, the decoy was C_A, who was about as attractive as A but with even a worse sense of humor. When confronted with this triad, the subjects preferred A to B. When the decoy was C_B, whose was about as comedic as B but much less attractive, subjects preferred B.[26] As with the hypothetical example of cars, one of the metrics of valuation was expanded by the decoy. In the first case it was humor and in the second case, attractiveness. Expanding a metric consequently devalued the male's phenotype who had excelled in that metric.

There is little evidence that we pick our cohorts to maximize competitive decoy effects. But we could. The previous section tells me that if I want to enhance my attractiveness, I should go out on the town with an attractive female in tow and let mate choice copying work its magic. If there are no attractive women around willing to help, we have just learned how I can do this with same-gender friends. I just have to be strategic about it. Let's say women tend to view my best buddy as favorably as they do me, but he is better looking and less funny. I should recruit a third friend to join us, and in choosing a recruit pay close attention to how his attractiveness and sense of humor relate to that of me and my best friend.

There is something very intuitive about mate choice copying, whereas competitive decoy effects just seem crazy. Is this a case where humans are overthinking things and where animals are actually "smarter," or at least more rational, than us? Or do they also fall prey to these decoys? We know far more about decoy effects in humans than in other animals.

Foraging is one domain in animal behavior where irrationality is known to rear its crazy head. It has been demonstrated in bees, hummingbirds, and Gray Jays. Let's take the Gray Jays as a case in point. In one experiment, jays were presented with a favorite food, raisins, at different distances inside a wire mesh funnel that the birds would have to traverse to get the food: two raisins at fifty-six centimeters versus one raisin at twenty-eight centimeters. We would predict that the jays would prefer more raisins and closer raisins; in this case, the jay's choices were biased toward the single, closer raisin. A competitive decoy was then introduced: two raisins at eighty-four centimeters. The birds succumbed to the decoy effect; they now preferred the two raisins at fifty-six centimeters.[27] Even though these effects seem crazy, at this point we should all be pretty good at predicting them. If you didn't correctly guess the jay's behavior yourself, go back and read the previous paragraphs again, at least before you go shopping for a car.

Can decoys influence the sexual aesthetics of animals? The only conclusive evidence comes from a study by Amanda Lea and me on túngara frogs. Amanda identified three mating calls whose relative attractiveness to females were documented years before. We refer to those measures as the call's "static attractiveness" because it is based on qualities of the call's sound that are fairly constant within a male and that elicit consistent preference rankings when they are presented to females in choice experiments at the same call rates, as explained in chapter 2. Females also prefer faster rather than slower call rates. Amanda combined the more attractive static call with the less attractive call rate and vice versa. When given a choice between two calls, A (higher static attractiveness, slower call rate) and B (lower static attractiveness, faster call rate), females had a slight preference for B. Amanda then introduced a decoy, call C. This call was similar to A in static attractiveness and had a substantially slower call rate than both A and B. Females preferred call A to call C, and B to C; call C, therefore, was an inferior alternative. When the females were given a choice between all three options, the preference switched from call B to call A.[28] This change in preference was predicted because, as with the cars, human attractiveness, and jays' raisins, the metric in which B was more attractive than A was call rate, and the introduction of C, with a much slower call rate than both A and B, expanded the range of call rates being compared—all of a sudden B's call rate did not seem that much better than A's. So we can't really answer the question: do female túngara frogs find call A or call B more attractive? It depends; theirs can be a fickle preference that depends on who else is singing along.

Decoys don't even need to be real; we now know that phantom decoys are lurking in the netherworld of many a marketplace. Let's go back to that car lot for a minute. What if, instead of showing you car

C, the salesperson merely described it to you, but said it was all sold out and she could not get you one: Would this still bias your choice to car A? Yes. Just knowing about a decoy is enough to skew you choice, thus the moniker *phantom decoy*. This makes it even easier to manipulate consumers, as a decoy in the bush has the same effect as a decoy in the hand. Phantom decoys also influence female túngara frogs. Amanda conducted a second experiment in which the speaker broadcasting the decoy was not on the floor, where females could approach it, but on the ceiling, where it was out of their reach—túngara frogs are not tree frogs and never climb to find a mate. The results were the same as before. Call C elicits a decoy effect if it is perceptible, even if it is not accessible.

Fickleness arising from decoys is well known in humans in several domains, including percepts of sexual beauty. My guess is that túngara frogs are hardly an exception in being influenced by sexual decoys, and decoy effects will account for a number of instances of fickle sexual preferences in the animal kingdom. Our studies also predict that courters in many species might have figured out how to use this effect to actively manipulate their attractiveness with the cunning and deceitfulness of a car salesperson.

This chapter has shown that biases in what we perceive as beautiful are not only properties of biases in our sensory systems and brains, but also biases brought about by the physiological and social contexts in which evaluations of beauty are made. All sources of bias also can give rise to preferences that remain hidden until some novel trait reveals it. I have mentioned hidden preferences throughout. In the next chapter, we will more thoroughly confront this important, but only recently appreciated, influence on the evolution of sexual beauty.

EIGHT

||||||||||||||||||||||||||||||||||||||||||||||||||||||||||||||||||||||||||||||||||||||||||

Hidden Preferences and Life in Pornotopia

There are known knowns . . . there are known unknowns. . . .
But there are also unknown unknowns—the ones we
don't know we don't know.   —*Donald Rumsfeld*

THIS WAS US SECRETARY OF DEFENSE DONALD RUMSFELD's response to a question about the lack of evidence for weapons of mass destruction in Iraq, evidence which the United States presented as a "known known" to justify invading that country. When the press dug deeper and found that no such evidence existed, the government's excuse was that it didn't know that it didn't know there were no weapons . . . or something along those lines.

Life is filled with discovering things we didn't know existed. This includes finding out that we sometimes like the unknowns. For 169 lines in Dr. Seuss's classic children's story *Green Eggs and Ham*, Sam explains in excruciating detail how he dislikes green eggs and ham, despite the fact that he has never tasted them: "I would not like them here or there. I would not like them anywhere. I do not like green eggs and ham. I do not like them, Sam-I-am."[1] Finally, after being coaxed to try them, Sam discovers that in fact he *does* like green eggs and ham. Sam had a hidden preference that stayed under the radar because it had never been probed by the right stimulus.

Traits of sexual beauty only evolve if they elicit a preference in a chooser. Thus, when we survey these traits in nature, we always find corresponding preferences for them in the opposite sex. This is hardly surprising, since beauty is in the brain of the beholder; beautiful traits only evolve if someone finds them beautiful. But how did this match between trait and preference, between the beauty of the bearer and the aesthetics of the beholder, come about?

Throughout much of this book, we have seen how preferences for traits can exist but remain untapped by courters until one of them just happens upon a trait, either through mutation or learning, that lures the hidden preference out into the open. A classic example is the experiment with swordtails and platyfish reviewed in chapter 3. When artificial swords were added to male platyfish (a group of fish that lack this trait), these newly endowed males were suddenly more attractive in the eyes of their females. This experiment seems to mimic the evolution of the sword in the swordtail; the preference was already in the platyfish-swordtail ancestor, perhaps as a general preference for larger males, which male swordtails exploited in an energetically cheaper way to appear larger.[2] Genetic mutations, rather than experimental manipulations, gave the male swordtails their sword. Since their females had a hidden preference for swords, the newly endowed males were immediately perceived as more attractive than their unendowed comrades.

In this chapter, I will delve deeper into the details of how traits and preferences come to match, and I will dwell a bit more on the relatively new hypothesis of hidden preferences and their exploitation. I briefly addressed this topic in chapter 3, but here I will consider the details and nuances of just how it happens.

* * *

Evolution can bring about the match between beautiful traits and the aesthetics that favor them in three ways: choosers can evolve preferences for already existing traits that deliver benefits to choosers; traits and preferences can evolve simultaneously; and, traits can evolve that are immediately found attractive because they exploit hidden preferences.[3] Let's look at each of these possibilities in detail.

Preferences can evolve when they enhance an individual's ability to reproduce, such as when they spark a desire for courters who are the correct species, fertile, disease-free, and who provide resources and care for offspring. These preferences will evolve to higher frequencies in a population than preferences that ignore these qualities of a potential mate, because these preferences result in choosers producing more offspring.

There are legions of examples of preferences that have evolved like this. Consider, for example, how feather color might evolve in birds. Imagine a population of Red-winged Blackbirds. Every spring, the marshes throughout North America echo the musical trill of males who sit on reeds as they flash their bright red sexual badges to females who are trying to decide on a partner. Then imagine that another species of blackbird, the Yellow-headed Blackbird, begins to cohabit the marsh, as they occasionally do. There are a few ways to tell these species apart, but for the female red-wingeds, the red wing-badge of the males is the most reliable. In this scenario, selection strongly favors female red-wingeds who mate only with males with red badges compared with females who show no preference for traits that distinguish between the males of the two blackbird species. These undiscriminating female Red-winged Blackbirds mate at random, and thus often mate with the wrong species. Furthermore, selection also favors females who prefer males with the brightest red badges because these are the males most different from the yellow-headed species. This type of open-ended preference for red is similar to that exhibited by the male zebra finches discussed in chapter 3, who prefer females who look the most different from males, females with the most-orange beaks. This is one way that preferences evolve to prefer certain traits—in this case, the traits that are more likely to ensure matings with males of the same species.

Now let's imagine that another visitor arrives; not another species of bird, but one of their parasites—feather lice. Some males have genes that make them resistant to the lice. Other males not endowed with genetic resistance are infected and become sickly, which in turn inhibits their ability to defend territories with high resources. These sickly males are not ideal mates, and this is a fact they cannot hide. Their infections quickly become public information as they are afflicted with something akin to a scarlet letter: the lice cause the male's red badge to fade from bright to dull. This results in another bout of selection that favors females choosing males with the brightest red badges, the healthiest males. These discriminating females now reap multiple benefits from their mate choice: fertilization by males who are the correct species; who have higher quality resources for the female and her offspring; avoiding males with sexually transmittable parasites; and obtaining genes for their offspring that might make them parasite-free.

In the case described above, female choice is based only on the male's sexual trait, the red badges; the females cannot see the male's genes for parasite resistance. Females gain the advantage of these parasite-resistant genes being passed on to their offspring because these genes are correlated with bright coloration. Therefore, bright coloration evolves under direct selection because it is directly favored by females, while parasite-resistant genes evolve by indirect selection because they are correlated with traits that are under direct selection, bright coloration. The anti-parasite genes hitchhike a ride across generations with the genes for bright coloration. In this case, female preferences can evolve for traits that benefit the females even if these are not the traits the females were targeting in their mate choice.

Not only can multiple male traits be correlated and thus evolve together; preferences and traits can also evolve simultaneously. Let's continue with this example, where badge brightness and genes for parasite resistance are correlated, and assume that a preference for bright males does not influence the female's immediate reproductive success. This is true in the wild, as female Red-winged Blackbirds almost always lay three or four eggs, irrespective of the father. But if females are able to channel these parasite-resistant genes, these good genes for survival, into their offspring, their babies will be more likely to reach adulthood. Since these offspring harbor not only parasite-resistant genes from their father, but also the genes to prefer red from their mother, both of these genes increase in frequency in the next generation. This is one way that preference and trait can evolve together. The preference evolves because it is linked to the survival genes and genes for red color, not because the preference is under direct selection. The logical case for the simultaneous evolution of good genes and preferences for "good genes" is strong, but the number of cases that support it is small relative to the research effort to find such effects over the past forty years.

Thanks to Ronald Fisher, we understand another way that indirect selection can happen. Quiz a statistician on Ronald Fisher, and she will tell you that he is among the greatest statisticians of the twentieth century and responsible for some fundamental contributions such as the analysis of variance and the eponymously named Fisher's exact test. Ask her about Fisher's contributions to evolution, and she is likely to draw a blank. Similarly, most evolutionary biologists are aware of Fisher's contributions to sex ratio theory, selection analysis, and his fundamental theorem of natural selection, but they often don't realize they have been using some of his statistical tools throughout their careers. Fisher was a man of great insights, and one of his most perceptive is Fisher's theory of runaway sexual selection.[4] This is sometimes called the sexy son hypothesis, and it is another way that preferences can evolve through indirect selection.

In the previous example, we saw how red badges and genes for parasite resistance can co-evolve. Runaway sexual selection proceeds in a similar fashion. The difference is that the female's preference for red becomes correlated with the male's "sexy" red badge. The greater the number of females that prefer red badges, the faster red badges will evolve in the population, and the faster will *preferences* for red badges also evolve, because female preference for red is correlated with the male's red badge. All those offspring from the sexy red-badged males have both red badge genes and red badge preference genes. This is another case of genetic hitchhiking, and it occurs without any increase in survivorship genes in the population.

As with preference evolution for "good genes," the logic underlying Fisher's runaway sexual selection hypothesis is strong, but only a few experimental studies support the notion that it has been an important force in the evolution of sexual beauty and the preferences for it. Fisher presented this idea in 1930 in his book *The Genetical Theory of Natural Selection*,[5] but it was not until another half-century had passed that Russ Lande and Mark Kirkpatrick verified Fisher's idea mathematically and set the stage for numerous studies searching for any footprint left behind in nature by Fisher's insight into the evolution of sexual beauty and the preferences for it.[6] There is now good evidence from studies on stalk-eyed flies that this process can occur in nature. This classic study shows that the genes for stalk length and for preferences for stalk length are inherited together when there is selection based on stalk length but no selection on the preference for it.[7]

We will now dive into the last process that can give rise to the match between sexual beauty and sexual aesthetics, one that occurs when courters evolve traits that exploit hidden preferences in choosers. Let's return to the blackbird scenario, but at a time before the males had evolved red badges. Occasionally, a mutation arises that causes the evolution of a red badge. As of yet there is no preference for males with these red badges, but there is a cost, as predators can spot these males more easily. All cost and no benefit quickly drives the mutation to extinction. Now, let's imagine that a new, supernutritious food source appears, a species of bright red worm that is more nutritious than the ubiquitous brown worms in the marsh. Selection now favors blackbirds who are better able to detect bright red worms. Subsequently, when a male evolves a red badge, he immediately attracts the attention of others who are now cued into red.

Being conspicuous is often the first step to getting a mate but also to becoming a meal. When the odds favor getting a mate, then the mutation for the conspicuous sexual trait should be favored in the population in spite of the risk it brings. The blackbird scenario with red worms is just an imaginary example, but real studies of real animals show this is not an unlikely scenario.

In the previous chapter, I talked about mate choice copying in guppies, and in that context, briefly mentioned that females prefer males with more orange. Females vary in how attractive they find orange. Different populations in the guppies' resident mountain streams of Trinidad show variation in both the strength of this preference by females and the amount of orange in males. As we might expect, the strength of the females' preference for orange and the amount of orange sported by the males are correlated among the river systems: in rivers where males have abundant orange coloration, females have strong preferences for orange; in rivers where the males are more dully colored, females have less of a preference for orange color. But what gives rise to variation in the preference? What is the foundation of the sexual aesthetic for orange?

Helen Rodd and her colleagues pointed out that guppies often feed on orange fruit.[8] The researchers suggested that the preference for food is at the root of the guppies' preference for orange males. The females are not fooled into thinking the males are fruits, but instead, the researchers hypothesized, the females develop a gestalt attraction to orange that spills over from their food preference into their mating preference. They tested this hypothesis by placing poker chips of different colors in tanks of male and female guppies from different populations whose females showed varying degrees of preference for orange coloration in males. Somewhat amazingly, the time inspecting orange chips by both sexes predicted the strength of females' preferences for orange courtship coloration in each population. The conclusion was that males evolved orange coloration to exploit a general attraction to orange that evolved in the domain of foraging. One could argue, however, that the direction of cause-and-effect might be reversed. Perhaps females initially evolved a preference for orange males that then predisposed them to prefer orange fruit. John Endler and Gemma Cole resolved this issue by recreating this evolutionary scenario in the lab.

Endler and Cole's approach was to artificially select for guppies to prefer certain colors of food and then ask if this would result in an evolutionary change in the male's color. They separated guppies into different groups, or "lines," and then presented them with a simulated food item that was either blue or red. The preference for food color evolved, and the subsequent generations of the two lines differed in their preferences for red versus blue food. Rodd's study would predict a correlated change in the female's preference for male color. And that is indeed what seemed to happen. As the food preference evolved across generations, so did the amount of orange sported by males: it increased in the lines that were selected to prefer red food and decreased in the lines selected to prefer blue food.[9] Of course, the males were constrained in their color change by available genes, so they could not turn actually red or blue, but orange and red stimulate very similar patterns of photoreceptors, while blue is quite different. The only obvious agent that could cause an increase in the male's coloration in this experiment was female preference. These experimental results seem to nail down the earlier interpretation by Rodd and her colleagues: preferences for orange fruits give rise to preferences for orange males.

When we observe sexual beauty and the preferences that favor it, we are seeing only the present, the mere tips of long branches on the tree of life that have been evolving for millennia. Without more information from careful experiments like those above, we cannot really glimpse the past processes that brought about this match. The arrow of causation between trait and preference can point either way, and in some cases even both ways. These three different evolutionary processes can reach the same endpoint for very different reasons.

* * *

There are no free lunches in the sexual marketplace. Regardless of how traits and preferences evolve, they incur costs as well as reap benefits. It is the cost-benefit ratio, and how this changes through time, that determines their legacy. Given the cost-benefit ratio involved in sensory exploitation of hidden preferences, this might be an especially easy process to trigger. Let me back up this assertion.

A hallmark of sexually attractive traits is that they are costly. Whether they are the showy tail of a peacock or the bright colors of a guppy, these traits usually take more energy to produce, more time to maintain, and are more conspicuous to predators than other types of traits. In the example of attractive red badges evolving in blackbirds, a mutation that causes red badges can quickly disappear from the population if it is attractive to predators but not yet attractive to females. This must happen often: a mutation gives rise to a conspicuous sexual trait, but it goes extinct while waiting around for a mutation in the preference gene that will deem this trait attractive and thus beneficial. If there are hidden preferences, however, then when a trait like the red badge arises, it will incur the same cost, but there is not the risk of waiting around for a preference mutation to occur—the benefit is immediately provided by the previously hidden but now-exposed preference. Thus, given the same mutation for an attractive trait, that trait is more likely to evolve if hidden preferences for it already exist.

Hidden preferences influence the evolution of sexual traits, but what causes the evolution of hidden preferences? There are numerous sources. They often arise from selection based on sensory, perceptual, and cognitive systems in other domains. Selection on color preferences related to food—such as the examples of the guppies, and the surf perch and bowerbirds discussed in chapter 4—show how selection on sensory systems in the domain of foraging results in hidden preferences for male courtship colors. Another possibility involves selection to correctly identify sex, which in zebra finches generates hidden preferences that can arise from peak shift displacement, as discussed in chapter 3. In most cases, we expect the origin of the hidden preferences to be an adaptive response to the world around them, with an influence on perceptions of beauty that is incidental rather than the immediate consequence of selection and evolution.

Hidden preferences will nearly always be related to adaptive advantages in other domains. Thus, to compute the evolutionary costs and benefits of a hidden preference, we must take into account not only how it impacts the chooser's mating success but also how it relates to functions in other domains that influence the chooser's fitness. Let's go back to the guppies and imagine that males with more orange attract more parasites, and by courting a more orange male rather than a less orange male, a female is more likely to become parasitized herself. If she gains no other benefits from the more orange male, we might assume that exhibiting this hidden preference for especially orange males is maladaptive, all costs and no benefits. And similar to a new sexual trait without the benefits of a preference, a newly exposed hidden preference with only costs should also go extinct. But if preferences for food and mate color are inextricably linked, then to give a fair accounting of the fitness costs and benefits of the hidden preference, we also need to take into account benefits of this bias-toward-orange in the foraging domain. The trait facing costs and benefits is not just "preference for oranger males" but "bias toward orange in general." This is reminiscent of the behavior of orchid bees mating with orchids I discussed in chapter 3. It seems downright silly and certainly maladaptive for an animal to have sex with a plant until we consider this perversion in the context of the bee's mate-searching strategy. Since female bees are hard to come by, then it is better for the male bee to be too eager to mate, and sometimes mate with a flower, rather than too discriminating, and sometimes pass up real female bees.

Yet tracing the evolution of courtship behavior in terms of traits that *exploit* hidden preferences might suggest that such inclinations are maladaptive. In fact, there are few, if any, examples of this being the case. Instead, the opposite usually holds: once hidden preferences are revealed, they deliver benefits rather than costs to the chooser. How is that? Hidden preferences probably reduce search costs to choosers.[10] Courters that exploit these preferences often do so because they are more conspicuous to choosers; for example, they are more easily seen in guppies, surf perch, and fiddler crabs, and more easily heard in many frogs, insects, and songbirds. In the fiddler crab example, males that erect towers by their burrows are more easily seen by females, since the structure of her eyes make her especially sensitive to objects protruding vertically from the surface. Besides being an extension of the male's sexual phenotype, the towers also direct females to shelter from predators.[11]

Not only are exploiting males more easily sensed; their sexual traits can facilitate quicker mating decisions and longer memories of the signals. Female túngara frogs decide on a mate faster when choosing between a whine-chuck and a whine than when faced with the choice between two whines. In addition, females also remember the location of a whine with multiple chucks over a whine-only or a whine with a single chuck.[12] My colleague Molly Cummings and I recently reviewed hundreds of cases in which males evolved sexual traits to exploit hidden preferences. In most cases, these preferences appear to aid rather than impede finding a mate, thus decreasing the time involved in searching.[13]

The sexual marketplace is a dangerous place, but one that can't be avoided. It is the only place to shop for a mate, but it is also filled with predators shopping for food and parasites looking for a home. The quicker the sexual consumer can get out of there, the less likely it will become the consumed. So having your hidden preferences exploited might not be all that bad—in fact it might be mostly good.

* * *

I have mentioned a few examples of traits that have evolved to exploit hidden preferences. What I find especially interesting are examples of hidden preferences for traits that are not only lacking in the species of interest but also in any closely related species, traits that researchers rather than evolution have brought into being. These examples give us insights into how the preference landscape is ripe with hidden preferences waiting to be exploited. It is this lability of the preference landscape that affords the chooser's brain so much creative power in driving the evolution of sexual beauty.

The ornithologist Nancy Burley conducted some early and insightful experiments on hidden preferences. When birds are kept in aviaries, it is hard for researchers to keep track of who is who. One solution is to put bands on their legs, and if the bands are different colors, then researchers can identify the birds without handling them. Zebra finches do not have leg bands in nature, so it was a shock when Burley found that leg bands influence the attractiveness of both sexes. Males are more attracted to females adorned with black and pink leg bands, and they are not attracted to females with light blue or light green leg bands. Females, on the other hand, prefer red-banded males and also avoid males with light blue or light green bands. Besides providing one of the early windows into hidden preferences, Burley's study was also important because it revealed that studies of mating success in captive birds could be biased by the use of leg bands. Burley also pushed the envelope a bit with her experiments. She adorned male grassfinches with what looked like "party hats." Some species of birds have crests, which are elongated feathers on the head, but there are 120 species of grassfinches, and none of them have crests. Yet when elongated feathers are added to the tops of males' heads in two grassfinch species, they look ridiculous to us but more sexually appealing to females than the typical males of their species.[14]

Other researchers have taken this approach of adding novel traits to males to search for hidden preferences in females. Mosquito fish have been introduced all over the world as a biocontrol agent. As its name implies, mosquito larvae can constitute a substantial component of the diets of these fish. In Australia, a country that is infamous for failed biocontrol efforts such as introducing cane toads, the introduced mosquito fish experiment failed because these fish outcompeted other natural mosquito predators. Like the cane toads, mosquito fish are now considered pests Down Under. There is nothing colorful or exciting about these fish. The males are small, only a few centimeters long, and they lack any conspicuous courtship traits or behaviors. Males have a sexual organ, a gonopodium, that they use to inseminate females, somewhat like a penis. But in most ways it is nothing like a penis. It is a long, modified fin with a groove on the outside. Sperm travels down this groove and, when the end is inserted into a female, the sperm enters the female's body. Except for the gonopodium, the males make little investment in sex; they lack the flashy colors of guppies and the sexual ornaments of swordtails. But . . . what if?

This is the question that the animal behaviorist Jim Gould and his colleagues asked. In twenty-nine separate experiments, they presented to females models of male mosquito fish that had been manipulated in a myriad of ways. Tail fins were stretched out, dorsal fins were morphed to appear sharklike, swords were added, fish were blackened, speckled, and whitewashed. In almost every comparison, females showed preferences for the weird, the novel, and the outlandish males.[15] The real males might be conservative in their approach to sexual beauty, but deep down, below the radar, their females are yearning for anything but conservative; they are teeming with hidden preferences. The same occurs in túngara frogs. Even though these males have outdone their close relatives by evolving the chuck, an incredibly attractive syllable that when added to their call increases a male's attractiveness by 500 percent, most of their females' acoustic desires are unmet. In a series of experiments quite similar to Gould's, we conducted thirty-one experiments in which we manipulated calls in varying ways, such as replacing the chuck with blasts of white noise, calls of other species, and even bells and whistles. Like Gould, we also found an astounding promiscuity of preferences.[16] Females found many of these acoustic accoutrements attractive, even bells and whistles. When the chuck evolved, it was lucky enough to exploit a hidden preference, but we now see it was not uniquely attractive. Many different kinds of sounds might have worked just as well; the luck of the chuck was being first.

The evolution of sexual beauty in many cases is analogous to an artist experimenting with paint on a canvas or a musician tinkering with new combinations of beats and chords. They are probing for what will ring true with the aesthetics of their audience. All three are creative processes. All of them surround us with beauty by probing deep into our brains to find out just what we will consider beautiful.

* * *

So what about us? Do we adopt sexual traits that exploit hidden sexual preferences? Of course we do, and we do so quite easily, especially because we can synthesize forms, images, and sexual scenarios. Industries that target our sexual aesthetics, much like the perfume industry, can create artificial stimuli, readily test them in the marketplace, and quickly determine those that happen to match preferences of consumers, whether those preferences are hidden or plain as day. I will end this chapter with two interesting examples of how commerce does this. One example is cute and entertaining; the other is disturbing.

First, the cute one—a toy doll that has become an icon in Western culture. I have six younger sisters and two daughters; for much of my life there has always been Barbie. Although Barbie is not a sex toy, some have argued she represents unrealistic standards for sexual beauty in women. Barbie is getting old, but she has hardly aged a day since her birthday, March 9, 1959, when she made her debut at the American International Toy Fair in New York City. Some folks have problems with Barbie, feeling that she promulgates a sexist view of women's place in society. But many consider her beautiful; she exudes cues of youth and fertility. Barbie is long and slender; her ample breasts indicate she is sexually mature; their perkiness testifies to her young age; and her long, luxurious hair is a signal of health. Some might think she is so beautiful that she is unreal, and they would be right. Barbie is a supernormal stimulus, exhibiting an exaggeration of traits that puts her in the realm of the unreal, as many have pointed out before—she is a fake.

Barbie is only one-sixth of life-size, so let's blow her up to normal size and see how she compares with the real thing. Some details of Barbie's physique are merely average. Her head circumference (twenty-two inches) is about normal, and her bust is only a bit on the smaller size (thirty-two versus thirty-five or thirty-six inches). But most of her other parts are dwarfed compared with those of an average woman. Petite is an understatement. Her waist (sixteen inches) and hips (twenty-nine inches) are tiny and give her a waist-to-hip ratio of 0.56, miniscule compared with real US women, who average around 0.80, and even much smaller than the 0.71 ratio that many men find more attractive than average, as discussed in chapter 7. Her neck, wrists, forearms, ankles—and especially her thighs—are like matchsticks. The life-size Barbie would be pretty useless in the real world. With her super-thin and extra-long neck, she wouldn't be able to raise her head; her tiny waist would limit her to half a liver and only a few inches of intestine; and with her tiny feet, thin ankles, and top-heavy form, she would have to walk on all fours.[17] Even though Barbie is dysfunctional, many still find her beautiful, a real doll! This might seem odd until we think about some parallels with real, live women.

According to *Forbes* magazine, Gisele Bündchen earned $42 million in 2013, making her the world's highest-paid supermodel at that time.[18] As the money proves, many people in Western society find supermodels superattractive. Of course, supermodels are not your typical Western women. The average supermodel is five feet, ten inches tall and 107 pounds, a good bit different from your average non-super US woman, who is five feet, four inches tall and 166 pounds. Supermodels exist, but they are rare. In the mass media, however, they are paraded around constantly so that we can all admire their beauty, purchase the products they are hawking, and be seduced to think that their beauty is normal. In fact, their type of beauty seems to be tugging at an otherwise hidden preference for Barbie-like bodies—one for abnormally long and slender women—a concealed predilection that might exist because of our biology, our culture, or some combination of the two.

Hidden preferences, like the hypothesized "Barbie" preference, exist under the radar of selection. As I have mentioned before, if the hidden preference is a detriment to the chooser once it is revealed, then it should be weeded out by selection. A preference for a Barbie-like woman would not have survived in preindustrial society, such as the "environment of evolutionary adaptedness" of the Pleistocene era when, as evolutionary psychologists argue, much of our current behaviors were formed.[19] Even if she happened to survive scurrying around on all fours and being limited to half a liver and barely any intestines, her birth canal would have been too narrow to pass a newborn. As Barbie went extinct, so would any preferences for her as a mate.

But we are not living in the Pleistocene, and hidden preferences for supernormal sexual stimuli no longer lurk under the radar. Today, a wide range of sexual stimuli are accessible with the click of a computer mouse. This simple act lures these hidden preferences out into the open and recruits these likings to support a $10 billion–dollar industry. Welcome to pornotopia.

* * *

Pornotopia, a current term to describe pornography in Victorian England, is a fantasy land of supernormal sexual stimuli created mostly for men.[20]

This fantasy land is peopled by women who exude sexuality. Typically they are young, seemingly just "barely legal," with long hair and legs, unblemished skin, full lips, and a waistline that has never been extended by a developing child. The women are real, although some of their parts might be artificial, but they are hardly normal, as they are drawn from the extremes of the distribution of what real women look like. But it is not just their looks that are extreme, their sexual behavior departs from the mean as well. As Catherine Salmon describes it, "Sex in pornotopia is all about lust and physical gratification, without courtship, commitment, mating effort or long-term relationships. In pornotopia, women are eager to have sex with strangers, easily sexually aroused and always orgasmic."[21] Pornotopia is the perfect place for men to put into practice that most basic of male mating strategies we encountered in chapter 1—think quantity not quality, mate often, and let the females make all the investments in offspring.

Excessive viewing of pornography is considered a compulsive sexual behavior, but according to the *DSM-5*, the latest version of the *Diagnostic and Statistical Manual of Mental Disorders*, it is not an addiction.[22] It is certainly a sexual fetish, which, as defined by L. F. Lowenstein in the journal *Sexuality and Disability*, is easy to know when you see it: "A sexual fetish is identified by the use of a nonliving object as the exclusive or preferred method of achieving sexual gratification."[23] We think of most animals as more utilitarian than we are; they reserve sex for its evolved function—reproduction. So I was quite surprised when I saw a psychologist at my own university, Michael Domjan, present a lecture on sexual conditioning in quails revealing that other animals can also develop sexual fetishisms.

Quail are good subjects for sexual research. They are easy and cheap to care for, they respond well in experiments, and they like to have sex. Most male birds lack a penis or any other type of intromittent organ, so sex involves a cloacal kiss instead of penile intromission. They have to climb onto the back of the female while the two sexes bring their cloacae into contact and the male "spits" a bit of sperm into her reproductive tract. Keep this bit of bird biology in mind as we explore the dark side of quail sex.

Domjan used Pavlovian conditioning to delve into the quails' own land of pornotopia.[24] We probably all need a reminder of how Pavlovian conditioning works; let's start with a joke: Pavlov walks into a bar; the bartender rings the bell over the counter signifying that it is time for the last round; and Pavlov exclaims, "I forgot to feed my dog!" If the joke does not reawaken the memories of Pavlov's classic classical-conditioning experiment, here is how it works. Typically, dogs salivate when they anticipate food. In his experiments, Pavlov rang a bell and then he gave the dog food. The dog salivated. Pavlov kept this up until the bell caused the dog to salivate in response to the bell in anticipation of the food. When this happened, the dog had been conditioned. In these types of experiments, the bell is the conditioned stimulus (CS), an artificial, experimental stimulus; the food is the natural, unconditioned stimulus (US); and the salivating response to the food is the natural, un-conditioned response (UR). The goal of the experiment is to condition the subject so the CS causes the UR—merely hearing the bell makes the dog salivate. Follow-up experiments can determine the strength of the CS-UR association once it is acquired by determining how long it takes for this association to be extinguished; for how many trials will the dog continue to salivate at the sound of the bell if the food never appears?

Back to the quail. Domjan and his colleagues placed a male quail in an arena. The male was presented with a terry cloth object placed over a vertical cylinder and filled with soft polyester fibers— something akin to a sex toy. This is the CS. The CS was presented for thirty seconds and then immediately followed by presentation of a live female, the US, for five minutes, which was usually plenty of time for the couple to mate. Conditioning took place when the male approached and inter-acted with the CS prior to the release of the female. There were thirty conditioning trials. This was then followed by thirty extinction trials in which the males had access to the sex toy but not to females.

These sexual-conditioning experiments were successful; after half-a-dozen or so trials, all of the males were conditioned; they exhibited the UR of approaching and inspecting the terry cloth object. The surprising result was that about half of the males actually tried to copulate with the inanimate object—they had developed a sexual fetish. This sex toy had little resemblance to a female except that it was soft; it had no cloaca-like opening that could welcome a male's cloacal kiss. Nevertheless, it still elicited copulation behavior from many of the males.

The conditioning trials were then followed by extinction trials. As before, males were presented with the sex toy, but this stimulus was never reinforced by the later presentation of live females. In these trials, most of the males eventually stopped interacting the terry cloth object, but the males who had developed a sexual fetish showed no decrease in their sexual appetite for the sexual toy. The terry cloth object now became a sexual object in its own right, or more precisely a sexual fetish. It was valued by the males not because it presaged an avenue for sexual gratification, a real female, but because it became an avenue for sexual gratification itself.[25] This series of experiments by Domjan and his colleagues did not uncover a hidden preference, like many of the experiments just discussed, but created a new preference, in this case a maladaptive one in which the preference for a fetish continues even when it is no longer associated with a real, live sexual partner.

The quail experiments did not investigate the neurochemical processes underlying the development of sexual fetishes. But these experiments offer a glimpse into how compulsive desires for pornography might develop in humans. In the case of humans, the underlying neurochemical processes are beginning to be explored.

We know something about what pornography does to the brain. In chapter 3, I discussed the difference between liking and wanting. The dopamine reward system is the one that causes us to want what we like. As we noted in that chapter, mice lick their whiskers when they like food. Block their dopamine receptors, and they still show the same degree of liking in response to a sugar treat as normal mice, but they are not willing to work to get more sugar. They like sugar but don't want it. Our brains are tuned for sex—we like it and want it.

A clever experiment with humans showed how liking and wanting can be disentangled when we view sexual beauty. Men were asked to rate computer images of faces of men and women in order of attractiveness. They were then allowed to view any of the faces that they pleased. They ranked facial attractiveness in both genders (liking) but then spent more time viewing attractive faces of women (wanting). The behavioral results were complemented by fMRI studies of brain activation; there was increased activation of brain areas associated with the dopamine reward system during the "wanting" compared to the "liking" activities of the test subjects.[26]

The dopamine reward system is an adaptive mechanism for animals to want things that are good for them in a Darwinian sense. Only in humans, it seems, has the reward system been exploited; gambling, eating, drugs, and sex have hijacked this system, leading to the demise of many an addict. Sex might be the activity that most easily exploits the reward system because, as J. R. Georgiadis pointed out in an article in *Socioaffective Neuroscience and Psychology*, sexual orgasm causes the most powerful natural dopaminergic reward in the human nervous system.[27] The potency of this positive reinforcement makes addiction to pornography easy to understand and easy to happen.

Both men and women view pornography, and they do so for what might be considered positive (for example, increase in sexual knowledge) and negative reasons (for example, interpersonal distress). Many studies show that men use pornography more often, are more drawn to hard-core pornography, and are more likely to be compulsive users of pornography.[28] Much of the research and discussion of compulsive pornography use addresses this problem in men, and this is where I will concentrate the discussion.

Many men like pornography because it is a supernormal stimulus, just as moths like the supernormal concentrations of sex pheromones and supernormal speeds of wing beats. While viewing pornography, men often masturbate and experience orgasm and with it an unparalleled charge of dopamine. This neurochemical charge cements the incentive salience of the pornographic images; it makes men not just like porn but want more of it. They develop a sexual fetishism attributable to a combination of the initial attraction to a supernormal stimulus and positive reinforcement owing to the orgasm and the resulting stimulation of the dopamine system. Liking leads to wanting; and in some, the wanting leads to compulsion; and despite what the *DSM-5* says, it seems like an addiction. In the extreme, these compulsions can lead to asocial or antisocial syndromes in which life in pornotopia replaces the real thing.

Not only can pornography become an object of one's sexual desires; it can also instruct how to act out these desires. Pornography is becoming one of the main conduits for sexual education. And consequently, it might shape neurons in our brain that teach us how to perform sex. Porn has replaced the locker room and health class as the disseminator of knowledge about the "birds and the bees." Prior to easy and widespread access to porn, few teenagers could receive firsthand tutoring about a wide assortment of sexual acts. The locker-room experts of days past, quite often only a few years older than their "students," might have known firsthand about kissing, petting, and getting to first, second, or even third base; but their knowledge had its limits. Not so with Internet porn, which is not only rife with a wealth of sexual activities, but provides graphic illustrations as to what they are and how to perform them, leaving little to the imagination.

In an essay in *Brain and Addiction*, Donald Hilton digs deep into this notion of pornography as a supernormal stimulus and raises another serious concern dealing with "mirror neurons."[29] Mirror neurons are visual-motor neurons that were first discovered in the prefrontal cortex of monkeys. These neurons fire both when a monkey performs a certain action and when it sees another performing the same action. One function of motor neurons is in aiding imitation. The pattern by which mirror neurons fire when observing an action can act as a template as to how they should fire when an individual performs the same motor pattern. A second function has to do with "action understanding." When motor neurons are fired by observing a particular action, the observer ascribes a meaning to that action based on what the observer would be doing to generate the same pattern of firing. When I watch someone swing a baseball bat, my mirror neurons that fire are the same ones that fire when I am swinging a bat. Thus, I now know what I am seeing.

In studies using fMRI of brain areas containing mirror neurons, subjects were asked to view video clips of pornography. The studies revealed increased neural activity during viewing that was correlated both with enhanced sexual emotions and penile erections. Although these studies demonstrate correlation and not causation, they do suggest a potential role of mirror neurons in either learning to imitate or learning the meaning of sexual acts. Given the increasingly violent and demeaning nature of some forms of pornography, Hilton is concerned about the potential "negative emotional, cultural, and demographic effects" that pornography stamps into various networks of the neural system that are involved in learning and in understanding how to appropriately interact with a sexual partner. A frightening consequence of this multibillion-dollar industry is that pornography might create neural templates redefining in our brains what is normal sexual behavior.

It must be noted that the function of mirror neurons is still a bit controversial, especially the function and even the existence of mirror neurons in humans.[30] But if pornography influences an individual's idea of what sex is supposed to be like, then Hilton's concern is still valid, regardless of whether mirror neurons are involved. This final section also suggests that Naomi Wolf was quite prescient in her assertion more than a decade ago that "for the first time in human history, the images' power and allure have supplanted that of real naked women. Today real naked women are just bad porn."[31]

Today, it now seems clear that supernormal stimuli, hidden preferences, and the neural circuits for liking and wanting all conspire to drive the pornography industry. The results are analogous to what has been happening in the evolution of sexual beauty for millennia. But instead of courters evolving traits that influence sexual preferences of choosers, humans have entire industries, including but not limited to pornography, invested in creating stimuli that target our sexual aesthetics in cultural rather than evolutionary time. Remember that the next time you hear birds sing, see a firefly flash, or a watch a supermodel hawking a product that you really don't need.

Epilogue

‖‖

BEAUTY IS ALL AROUND US, AND IT IS INTOXICATINGLY DIVERSE. Much of this diversity exists because beauty enters our sexual brain through different sensory modalities, which strains our ability to make comparisons: we cannot objectively rank the beauty of a dance, a song, and a fragrance. The diversity of beauty is no less astounding within a single sensory domain—the collage of colors of many fishes and the vocal repertoires of songbirds are both overwhelming. The existence of all this diversity makes it obvious that there is no single Platonic ideal of beauty. This is true within our own species and also among the hundreds of thousands of species that reproduce sexually. The diversity of beauty springs from the diversity in how different species and even individuals of the same species sense the world around them. Our sexual aesthetics, those of humans and other species, are not handed down from above but are generated from within, specifically from within our brains. We are the ones who define beauty, and understanding the existence of beauty as well as our taste for it is not possible without viewing beauty through the brain of the beholder. If nothing else, I hope I have convinced you of this fact.

The brain sciences, which embrace neuroscience, psychology, and several areas of medicine, are making amazing strides in the beginnings of what has been called The New Century of the Brain. Often the relation between the brain and evolution is an afterthought. When researchers do consider the two in tandem, the focus is usually on how the brain evolved to be as it is. Regardless of the target species, this is a compelling question, but an equally compelling one is how the brain drives evolution. This book presents one scenario of how this happens.

I have considered how the brain drives the evolution of beauty. But I have done so mostly in one scenario, when courters evaluate choosers during mate choice in heterosexual pairs. But there is certainly more diversity in sexual behavior than I have covered here. Most of my examples involve females choosing males, or mutual evaluation by males and females. Although I have mentioned males choosing females, I have not drilled deeply into the factors that flip this equation from the more typical form of females choosing males. Biologists understand why this happens; it is just not within the scope of this book.

I also have not considered homosexual pairs, a phenomenon that is not restricted to humans. There are many compelling questions here, but we might be asking the wrong questions if we consider heterosexual/ homosexual as two invariant categories rather than two ends of a spectrum of sexual preference. Nevertheless, it would be of interest to know if "homosexual" individuals evaluate beauty in same-gender individuals by using the very same parameters that members of the opposite gender would employ to evaluate the beauty of the same individuals. And, if not, why not? Understanding the evolution of sexual beauty in a heterosexual mating paradigm is an important quest, but not the only one.

Of course, beauty is not restricted to sexual beauty. The perspective I present here also leads us to wonder how the idiosyncrasies and quirks in our own brains influence our own appreciation of "beauty" in a grander sense, as it applies beyond sex. Why is a rainbow "beautiful"? Why does the mere refraction of light into bands of color inspire awe? We can pose the same question about a work of art, a field of flowers, and an expertly executed move on the football field. Might any of these percepts of beauty be a side effect of our sexual aesthetics? Or, alternatively, can our appreciation of beauty in other domains influence what we find to be sexually beautiful? What is it about our senses, our brains, and our cognitive architecture that gives us an appreciation for beauty all around us? Why does beauty matter so much?

Like Darwin, we will continue to be confounded by aspects and elements of beauty as we encounter them, but we have come to understand so much more about the evolution of beauty since his time. As scientific exploration continues into the future, we are sure to expand even further our ability to see the ways that beauty is woven into existence, the many forms it assumes, and the wild appreciations it elicits.

Notes

||||||||||||||||||||||||||||||||||||||||||||||||||||||||||||||||||||||||||||||||||||||||||||||||||

Chapter 1. Why All the Fuss about Sex?

Epigraph: Darwin (1860).
1. Ryan (2010).
2. Darwin (1859).
3. Malthus (1798).
4. Slotten (2004).
5. Smith (1990).
6. Darwin (1871).
7. Diamond (1992).
8. Moen, Pastor, and Cohen (1999).
9. Emlen (2014).
10. Yoshizawa, Ferreira, Kamimura, and Lienhard (2014).
11. Yeung, Anapolski, Depenbusch, Zitzmann, and Cooper (2003).

Chapter 2. Why All the Whining and Chucking?

Epigraph: Song quoted in Langstaff and Rojankovsky (1955).
1. Simpson (1980).
2. McCullough (2001).
3. Ryan (2006).
4. Ryan (1985; 2011).
5. Collins (2000).
6. Evans, Neave, and Wakelin (2006).
7. Buss (1994).
8. Zoological Society of London, https://www.zsl.org/cheetah-fast-facts.
9. Tuttle (2015).
10. Griffin (1958).
11. Bruns, Burda, and Ryan (1989).
12. Johnston, Hagel, Franklin, Fink, and Grammer (2001).
13. Petrie and Williams (1993).
14. Capranica (1965).
15. Hoke, Burmeister, Fernald, Rand, Ryan, and Wilczynski (2004).
16. Wilczynski, Rand, and Ryan (2001).
17. Ryan (1990).

Chapter 3. Beauty and the Brain

Epigraph: Hume (1742).
    1. Von Uexküll (2014).
    2. Internet Archive, https://archive.org/details/drac_stoker.
    3. Galambos (1942).
    4. Griffin (1958).
    5. Nagel (1974).
    6. Feng, Narins, Xu, Lin, Yu, Qiu, Xu, and Shen (2006).
    7. Kurtovic, Widmer, and Dickson (2007).
    8. Taylor and Ryan (2013).
    9. Toda, Zhao, and Dickson (2012).
    10. Meierjohann and Schartl (2006).
    11. Basolo (1990).
    12. Jersáková, Johnson, and Kindlmann (2006).
    13. Zahavi (1975); Zahavi and Zahavi (1997).
    14. Silver (2012).
    15. Searcy (1992).
    16. ten Cate and Rowe (2007).
    17. ten Cate, Verzijden, and Etman (2006).
    18. Ryan and Keddy-Hector (1992).
    19. Weber (1978).
    20. Cohen (1984).
    21. Akre, Farris, Lea, Page, and Ryan (2011).
    22. Heath and Mickle (1960).
    23. Kringelbach and Berridge (2012).
    24. US Food and Drug Administration, http://www.fda.gov/NewsEvents/Newsroom/PressAnnouncements/ucm458734.htm.

Chapter 4. Visions of Beauty

Epigraph: Emerson (1899).
    1. Written by Pete Townshend, performed by The Who, "See Me, Feel Me," *Tommy* (1969).
    2. Dunn, Halenar, Davies, Cristobal-Azkarate, Reby, Sykes, Dengg, Fitch, and Knapp (2015).
    3. Quoted in *New World Encyclopedia*, http://www.newworldencyclopedia.org/entry/Howler_monkey.
    4. Dominy and Lucas (2001).
    5. Darwin (1872).
    6. Written by Lou Reed, performed by the Velvet Underground, "Sweet Jane," *Loaded* (1970).
    7. Changizi (2010).
    8. Ewert (1987).
    9. Hubel and Wiesel (1962).
    10. Rothenberg (2012).

11. Cummings (2007).
12. Magnus (1958).
13. Tuttle (2015).
14. Andersson (1994).
15. Andersson (1982).
16. Møller and Thornhill (1998).
17. Møller (1992).
18. Ryan, Warkentin, McClelland, and Wilczynski (1995).
19. Ghirlanda, Jansson, and Enquist (2002).
20. Phelps and Ryan (1998).
21. Enquist and Arak (1994).
22. Møller and Swaddle (1997).
23. TED Talk, https://www.ted.com/talks/cameron_russell_looks_aren_t_every thing_believe_me_i_m_a_model?language=en.
24. Dawkins (2006).
25. Dawkins (1999).
26. Slotten (2004).
27. Diamond (1999).
28. Diamond (1992).
29. Madden and Tanner (2003).
30. Kelley and Endler (2012).
31. Chatterjee (2011).

Chapter 5. The Sounds of Sex

Epigraph: Marler (1998).
1. Tom Harrington, "About Deafness," under "FAQ: Deaf People in History; Quotes by Helen Keller," Gallaudet University Library, February 2000, http://libguides .gallaudet.edu/content.php?pid=352126&sid=2881882.
2. Carson (1962).
3. Rodríguez-Brenes, Rodriguez, Ibáñez, and Ryan (2016).
4. Rodríguez-Brenes, Garza, and Ryan (unpublished data).
5. O'Connor, Fraccaro, Pisanski, Tigue, O'Donnell, and Feinberg (2014).
6. Zuk, Rotenberry, and Tinghitella (2006).
7. Pascoal, Cezard, Eik-Nes, Gharbi, Majewska, Payne, Ritchie, Zuk, and Bailey (2014).
8. O'Connor, Fraccaro, Pisanski, Tigue, O'Donnell, and Feinberg (2014).
9. Morton (1975).
10. Hunter and Krebs (1979); Ryan, Cocroft, and Wilczynski (1990).
11. Halfwerk, Bot, Buikx, van der Velde, Komdeur, ten Cate, and Slabbekoorn (2011).
12. Hartshorne (1973).
13. Searcy (1992).
14. Mello, Nottebohm, and Clayton (1995).
15. Pfaff, Zanette, MacDougall-Shackleton, and MacDougall-Shackleton (2007).
16. Lehrman (1965).

17. Cheng (2008).
18. Earp and Maney (2012).
19. Wyttenbach, May, and Hoy (1996).
20. Nakano, Takanashi, Skals, Surlykke, and Ishikawa (2010).
21. Proctor (1992).
22. Cui, Tang, and Narins (2012).
23. Lardner and bin Lakim (2002).
24. Clark and Feo (2008).
25. Bostwick and Prum (2005).
26. Morton (1977).
27. McConnell (1990).
28. Juslin and Västfjäll (2008).
29. Steblin (2002).
30. Mitchell, DiBartolo, Brown, and Barlow (1998).
31. Blood and Zatorre (2001).
32. Menon and Levitin (2005).

Chapter 6. The Aroma of Adulation

Epigraph: Helen Keller, *The World I Live In*, chap. 6, "Smell: The Fallen Angel," reprinted in *Ragged Edge Online*, 5 (September 2001): http://www.raggededgemagazine.com /0901/0901ft3-2.htm.
1. Grosjean, Rytz, Farine, Abuin, Cortot, Jefferis, and Benton (2011).
2. Seeley (2009).
3. Prosen, Jaeger, and Lee (2004).
4. Bradbury and Vehrencamp (2011).
5. Domingue, Haynes, Todd, and Baker (2009).
6. Ibid.
7. Ryan and Rosenthal (2001).
8. Written and performed by Janis Ian, "Society's Child" (1965), *Between the Lines* (1975).
9. Fisher, Wong, and Rosenthal (2006).
10. Meyer, Kircher, Gansauge, Li, Racimo, Mallick, Schraiber, et al. (2012).
11. McClintock (1971).
12. Miller (2011).
13. Wedekind, Seebeck, Bettens, and Paepke (1995).
14. Garver-Apgar, Gangestad, Thornhill, Miller, and Olp (2006).
15. Villinger and Waldman (2008).
16. Vollrath and Milinski (1995).
17. Rodríguez-Brenes, Rodriguez, Ibáñez, and Ryan (2016).
18. Schiestl (2005).
19. Burr (2004).
20. Milinski (2006).
21. Milinski (2003).
22. Milinski and Wedekind (2001).

Chapter 7. Fickle Preferences

Epigraph: Virgil, *Aeneid*, 4.569, in *Bartlett's Familiar Quotations*, https://books.google .com/books?id=W3SG1hJSArIC&pg=RA1-PR39&lpg=RA1-PR39&dq=virgil+ aeneid+A+woman+is+always+a+fickle,+unstable+thing.

1. Written by Baker Knight, performed by Mickey Gilley, "Don't the Girls All Get Prettier at Closin' Time," *Gilley's Smokin'* (1976).

2. Pennebaker, Dyer, Caulkins, Litowitz, Ackreman, Anderson, and McGraw (1979).

3. Johnco, Wheeler, and Taylor (2010).

4. Trivers (2011).

5. Haselton, Mortezaie, Pillsworth, Bleske-Rechek, and Frederick (2007).

6. Bryant and Haselton (2009).

7. Wyrobek, Eskenazi, Young, Arnheim, Tiemann-Boege, Jabs, Glaser, Pearson, and Evenson (2006).

8. Easton, Confer, Goetz, and Buss (2010).

9. Lynch, Rand, Ryan, and Wilczynski (2005).

10. Partridge and Farquhar (1981).

11. Lone, Venkataraman, Srivastava, Potdar, and Sharma (2015).

12. Lin, Cao, Sethi, Zeng, Chin, Chakraborty, Shepherd, et al. (2016).

13. Wiley (1973).

14. Dugatkin (1992).

15. Schlupp, Marler, and Ryan (1994).

16. Hill and Ryan (2006).

17. Henrich, Heine, and Norenzayan (2010).

18. Sugiyama (2004).

19. Sigall and Landy (1973).

20. Waynforth (2007).

21. Hill and Buss (2008).

22. Jarod Kintz, *This Book Is Not for Sale*, Amazon Digital Services, Kindle ed., May 2011, http://www.amazon.com/This-Book-SALE-Jarod-Kintz-ebook/dp/B0051OE DDA.

23. Winegard, Winegard, and Geary (2013).

24. Kirkpatrick, Rand, and Ryan (2006).

25. Courtiol, Raymond, Godelle, and Ferdy (2010).

26. Sedikides, Ariely, and Olsen (1999).

27. Shafir, Waite, and Smith (2002).

28. Lea and Ryan (2015).

Chapter 8. Hidden Preferences and Life in Pornotopia

Epigraph: US Department of Defense, "DoD News Briefing—Secretary Rumsfeld and Gen. Myers," February 12, 2002, http://archive.defense.gov/Transcripts/Transcript.aspx ?TranscriptID=2636.

1. Seuss (1988).

2. Rosenthal and Evans (1998).

3. Kirkpatrick and Ryan (1991).

4. Fisher (1930).

5. Ibid.

6. Lande (1981); Kirkpatrick (1982).

7. Wilkinson and Reillo (1994).

8. Rodd, Hughes, Grether, and Baril (2002).

9. John Endler, personal communication.

10. Ryan and Cummings (2013).

11. Christy and Salmon (1991).

12. Ryan, unpublished data.

13. Ryan and Cummings (2013).

14. Burley and Symanski (1998).

15. Gould, Elliott, Masters, and Mukerji (1999).

16. Ryan, Bernal, and Rand (2010).

17. Samantha Olson, "Barbie's Body Measurements Set Unrealistic Goals for Little Girls: Sales Plummet," Medical Daily, December 31, 2014, http://www.medicaldaily.com/pulse/barbies-body-measurements-set-unrealistic-goals-little-girls-sales-plummet-316006.

18. *Forbes*, http://www.forbes.com/pictures/eimi45mdj/no-1-gisele-bndchen/#7ed42a453c02.

19. Prescott (2012).

20. Marcus (2008).

21. Salmon (2012).

22. American Psychiatric Association (2013).

23. Lowenstein (2002).

24. Köksal, Domjan, Kurt, Sertel, Örüng, Bowers, and Kumru (2004).

25. Ibid.

26. Aharon, Etcoff, Ariely, Chabris, O'Connor, and Breiter (2001).

27. Georgiadis (2012).

28. Hald (2006).

29. Hilton (2013).

30. Turella, Pierno, Tubaldi, and Castiello (2009).

31. Wolf (2003).

Bibliography

||||||||||||||||||||||||||||||||||||||||||||||||||||||||||||||||||||||||||||||||||||||||||||

Aharon, I., Etcoff, N., Ariely, D., Chabris, C. F., O'Connor, E., and Breiter, H. C. (2001). Beautiful faces have variable reward value: fMRI and behavioral evidence. *Neuron* 32: 537–51.

Akre, K. L., Farris, H. E., Lea, A. M., Page, R. A., and Ryan, M. J. (2011). Signal perception in frogs and bats and the evolution of mating signals. *Science* 333: 751–52.

American Psychiatric Association (2013). *Diagnostic and Statistical Manual of Mental Disorders (DSM-5)*. Washington, DC: American Psychiatric Association Publishing.

Andersson, M. (1982). Female choice selects for extreme tail length in a widowbird. *Nature* 299: 818–820.

——— (1994). *Sexual Selection*. Princeton, NJ: Princeton University Press.

Basolo, A. L. (1990). Female preference predates the evolution of the sword in swordtail fish. *Science* 250: 808–10.

Blood, A. J., and Zatorre, R. J. (2001). Intensely pleasurable responses to music correlate with activity in brain regions implicated in reward and emotion. *Proceedings of the National Academy of Sciences of the United States of America* 98: 818–23.

Bostwick, K. S., and Prum, R. O. (2005). Courting bird sings with stridulating wing feathers. *Science* 309: 736.

Bradbury, J. W., and Vehrencamp, S. L. (2011). *Principles of Animal Communication*. Sunderland, MA: Sinauer Associates.

Bruns, V., Burda, H., and Ryan, M. J. (1989). Ear morphology of the frog-eating bat (*Trachops cirrhosus*, family: Phyllostomidae): Apparent specializations for low-frequency hearing. *Journal of Morphology* 199: 103–18.

Bryant, G. A., and Haselton, M. G. (2009). Vocal cues of ovulation in human females. *Biology Letters* 5: 12–15.

Burley, N. T., and Symanski, R. (1998). "A taste for the beautiful": Latent aesthetic mate preferences for white crests in two species of Australian grassfinches. *American Naturalist* 152: 792–802.

Burr, C. (2004). *The Emperor of Scent: A True Story of Perfume and Obsession*. New York: Random House.

Buss, D. M. (1994). *The Evolution of Desire*. New York: Basic Books.

Capranica, R. R. (1965). *The Evoked Vocal Response of the Bullfrog*. MIT Press Research Monograph, no. 33. Cambridge, MA: MIT Press.

Carson, R. (1962). *Silent Spring* Greenwich, CT: Fawcett Publications.

Changizi, M. (2010). *The Vision Revolution: How the Latest Research Overturns Everything We Thought We Knew about Human Vision*. Dallas, TX: Benbella Books.

Chatterjee, A. (2011). Neuroaesthetics: A coming of age story. *Journal of Cognitive Neuroscience* 23: 53–62.

Cheng, M.-F. (2008). The role of vocal self-stimulation in female responses to males: Implications for state-reading. *Hormones and Behavior* 53: 1–10.

Christy, J. H., and Salmon, M. (1991). Comparative studies of reproductive behavior in mantis shrimps and fiddler crabs. *American Zoologist* 31: 329–37.

Clark, C. J., and Feo, T. J. (2008). The Anna's hummingbird chirps with its tail: A new mechanism of sonation in birds. *Proceedings of the Royal Society of London B: Biological Sciences* 275: 955–62.

Cohen, J. (1984). Sexual selection and the psychophysics of female choice. *Zeitschrift für Tierpsychologie* 64: 1–8.

Collins, S. A. (2000). Men's voices and women's choices. *Animal Behaviour* 60: 773–80.

Courtiol, A., Raymond, M., Godelle, B., and Ferdy, J. B. (2010). Mate choice and human stature: Homogamy as a unified framework for understanding mating preferences. *Evolution* 64: 2189–203.

Cui, J., Tang, Y., and Narins, P. M. (2012). Real estate ads in Emei music frog vocalizations: Female preference for calls emanating from burrows. *Biology Letters* 8: 337–40.

Cummings, M. E. (2007). Sensory trade-offs predict signal divergence in surfperch. *Evolution* 61: 530–45.

Darwin, C. (1859). *On the Origin of Species*. London: J. Murray.

——— (1860). Charles Darwin to Asa Gray, April 3. Darwin Correspondence Project, Cambridge University. http://www.darwinproject.ac.uk/letter/?docId=letters/DCP-LETT-2743.xml;query=2743;brand=default.

——— (1871). *The Descent of Man and Selection in Relation to Sex*. London: J. Murray.

——— (1872). *The Expression of the Emotions in Man and Animals*. London: J. Murray.

Dawkins, R. (1999). *The Extended Phenotype: The Long Reach of the Gene*. Oxford: Oxford Paperbacks.

——— (2006). *The Selfish Gene*. Oxford: Oxford University Press.

Diamond, J. (1992). *The Third Chimpanzee*. New York: HarperCollins.

——— (1999). *Guns, Germs, and Steel: The Fates of Human Societies*. New York: W. W. Norton.

Domingue, M. J., Haynes, K. F., Todd, J. L., and Baker, T. C. (2009). Altered olfactory receptor neuron responsiveness is correlated with a shift in behavioral response in an evolved colony of the cabbage looper moth, *Trichoplusia ni. Journal of Chemical Ecology* 35: 405–15.

Dominy, N. J., and Lucas, P. W. (2001). Ecological importance of trichromatic vision to primates. *Nature* 410: 363–66.

Dugatkin, L. A. (1992). Sexual selection and imitation: Females copy the mate choice of others. *American Naturalist* 139: 1384–89.

Dunn, J. C., Halenar, L. B., Davies, T. G., Cristobal-Azkarate, J., Reby, D., Sykes, D., Dengg, S., Fitch, W. T., and Knapp, L. A. (2015). Evolutionary trade-off between vocal tract and testes dimensions in howler monkeys. *Current Biology* 25: 2839–44.

Earp, S. E., and Maney, D. L. (2012). Birdsong: Is it music to their ears? *Frontiers in Evolutionary Neuroscience* 4: 14.

Easton, J. A., Confer, J. C., Goetz, C. D., and Buss, D. M. (2010). Reproduction expediting: Sexual motivations, fantasies, and the ticking biological clock. *Personality and Individual Differences* 49: 516–20.

Emerson, R. W. (1899). *The Early Poems of Ralph Waldo Emerson*: T. Y. Crowell and Co. Google Books. https://books.google.com/books?hl=en&lr=&id=YFARAAAA YAAJ&oi=fnd&pg=PA1&dq=If+eyes+were+made+for+seeing,+Then+Beauty+is+

its+own+excuse+for+being.+Emerson+1899+&ots=X7se7ZSdQv&sig=K-hrqu vmuqY8wRdkZr2qKe4ZTFU#v=onepage&q&f=false.

Emlen, D. J. (2014). *Animal Weapons: The Evolution of Battle*. New York: Henry Holt.

Enquist, M., and Arak, A. (1994). Symmetry, beauty and evolution. *Nature* 372: 169–70.

Evans, S., Neave, N., and Wakelin, D. (2006). Relationships between vocal characteristics and body size and shape in human males: An evolutionary explanation for a deep male voice. *Biological Psychology* 72: 160–63.

Ewert, J.-P. (1987). Neuroethology of releasing mechanisms: Prey-catching in toads. *Behavioral and Brain Sciences* 10: 337–68.

Feng, A. S., Narins, P. M., Xu, C.-H., Lin, W.-Y., Yu, Z.-L., Qiu, Q., Xu, Z.-M., and Shen, J.-X. (2006). Ultrasonic communication in frogs. *Nature* 440: 333–36.

Fisher, H. S., Wong, B. B., and Rosenthal, G. G. (2006). Alteration of the chemical environment disrupts communication in a freshwater fish. *Proceedings of the Royal Society of London B: Biological Sciences* 273: 1187–93.

Fisher, R. A. (1930). *The Genetical Theory of Natural Selection*. Oxford: Oxford University Press.

Galambos, R. (1942). The avoidance of obstacles by flying bats: Spallanzani's ideas (1794) and later theories. *Isis* 34: 132–40.

Garver-Apgar, C. E., Gangestad, S. W., Thornhill, R., Miller, R. D., and Olp, J. J. (2006). Major histocompatibility complex alleles, sexual responsivity, and unfaithfulness in romantic couples. *Psychological Science* 17: 830–35.

Georgiadis, J. R. (2012). Doing it . . . wild? On the role of the cerebral cortex in human sexual activity. *Socioaffective Neuroscience and Psychology* 2: 17,337. doi: 10.3402/snp .v2i0.17337.

Ghirlanda, S., Jansson, L., and Enquist, M. (2002). Chickens prefer beautiful humans. *Human Nature* 13: 383–89.

Gould, J. L., Elliott, S. L., Masters, C. M., and Mukerji, J. (1999). Female preferences in a fish genus without female mate choice. *Current Biology* 9: 497–500.

Griffin, D. (1958). *Listening in the Dark: The Acoustic Orientation of Bats and Men*. New Haven, CT: Yale University Press.

Grosjean, Y., Rytz, R., Farine, J.-P., Abuin, L., Cortot, J., Jefferis, G. S., and Benton, R. (2011). An olfactory receptor for food-derived odours promotes male courtship in *Drosophila*. *Nature* 478: 236–40.

Hald, G. M. (2006). Gender differences in pornography consumption among young heterosexual Danish adults. *Archives of Sexual Behavior* 35: 577–85.

Halfwerk, W., Bot, S., Buikx, J., van der Velde, M., Komdeur, J., ten Cate, C., and Slabbekoorn, H. (2011). Low-frequency songs lose their potency in noisy urban conditions. *Proceedings of the National Academy of Sciences of the United States of America* 108: 549–54.

Hartshorne, C. (1973). *Born to Sing*. Bloomington: Indiana University Press.

Haselton, M. G., Mortezaie, M., Pillsworth, E. G., Bleske-Rechek, A., and Frederick, D. A. (2007). Ovulatory shifts in human female ornamentation: Near ovulation, women dress to impress. *Hormones and Behavior* 51: 40–45.

Heath, R. G., and Mickle, W. A. (1960). Evaluation of seven years' experience with depth electrode studies in human patients. In Ramey, E. R., and O'Doherty, D. eds., *Electrical Studies of the Unanesthetized Brain*. New York: Paul B. Hoeber.

Henrich, J., Heine, S., and Norenzayan, A. (2010). The weirdest people in the world? *Behavioral and Brain Sciences* 33: 61–83.

Hill, S. E., and Buss, D. M. (2008). The mere presence of opposite-sex others on judgments of sexual and romantic desirability: Opposite effects for men and women. *Personality and Social Psychology Bulletin* 34: 635–47.

Hill, S. E., and Ryan, M. J. (2006). The role of model female quality in the mate choice copying behaviour of sailfin mollies. *Biology Letters* 2: 203–5.

Hilton, D. L. (2013). Pornography addiction—a supranormal stimulus considered in the context of neuroplasticity. *Socioaffective Neuroscience and Psychology* 3: 20,767. doi: 10.3402/snp.v3i0.20767.

Hoke, K. L., Burmeister, S. S., Fernald, R. D., Rand, A. S., Ryan, M. J., and Wilczynski, W. (2004). Functional mapping of the auditory midbrain during mate call reception. *Journal of Neuroscience* 24: 11,264–72.

Hubel, D. H., and Wiesel, T. N. (1962). Receptive fields, binocular interaction and functional architecture in the cat's visual cortex. *Journal of Physiology* 160: 106–54.

Hume, D. (1742). *David Hume's Essays, Moral and Political, 1742*. Phrase Finder. http://www.phrases.org.uk/meanings/beauty-is-in-the-eye-of-the-beholder.html.

Hunter, M. L., and Krebs, J. R. (1979). Geographical variation in the song of the great tit (*Parus major*) in relation to ecological factors. *Journal of Animal Ecology* 48: 759–85.

Jersáková, J., Johnson, S. D., and Kindlmann, P. (2006). Mechanisms and evolution of deceptive pollination in orchids. *Biological Reviews* 81: 219–35.

Johnco, C., Wheeler, L., and Taylor, A. (2010). They do get prettier at closing time: A repeated measures study of the closing-time effect and alcohol. *Social Influence* 5: 261–71.

Johnston, V. S., Hagel, R., Franklin, M., Fink, B., and Grammer, K. (2001). Male facial attractiveness: Evidence for hormone-mediated adaptive design. *Evolution and Human Behavior* 22: 251–67.

Juslin, P. N., and Västfjäll, D. (2008). Emotional responses to music: The need to consider underlying mechanisms. *Behavioral and Brain Sciences* 31: 559–75.

Kelley, L. A., and Endler, J. A. (2012). Illusions promote mating success in great bowerbirds. *Science* 335: 335–38.

Kirkpatrick, M. (1982). Sexual selection and the evolution of female choice. *Evolution* 36: 1–12.

Kirkpatrick, M., Rand, A. S., and Ryan, M. J. (2006). Mate choice rules in animals. *Animal Behaviour* 71: 1215–25.

Kirkpatrick, M., and Ryan, M. J. (1991). The paradox of the lek and the evolution of mating preferences. *Nature* 350: 33–38.

Köksal, F., Domjan, M., Kurt, A., Sertel, Ö., Örüng, S., Bowers, R., and Kumru, G. (2004). An animal model of fetishism. *Behaviour Research and Therapy* 42: 1421–34.

Kringelbach, M. L., and Berridge, K. C. (2012). The joyful mind. *Scientific American* 307: 40–45.

Kurtovic, A., Widmer, A., and Dickson, B. J. (2007). A single class of olfactory neurons mediates behavioural responses to a *Drosophila* sex pheromone. *Nature* 446: 542–46.

Lande, R. (1981). Models of speciation by sexual selection on polygenic traits. *Proceedings of the National Academy of Sciences of the United States of America* 78: 3721–25.

Langstaff, J. M., and Rojankovsky, F. (1955). *Frog Went A-Courtin'*. Boston: Houghton Mifflin Harcourt.

Lardner, B., and bin Lakim, M. (2002). Animal communication: Tree-hole frogs exploit resonance effects. *Nature* 420: 475.

Lea, A. M., and Ryan, M. J. (2015). Irrationality in mate choice revealed by túngara frogs. *Science* 349: 964–66.

Lehrman, D. S. (1965). Interaction between internal and external environments in the regulation of the reproductive cycle of the ring dove. In Beach, F. A., ed., *Sex and Behavior*, 355–80. New York: Wiley.

Levitin, D. J. (2011). *This Is Your Brain on Music: Understanding a Human Obsession.* London: Atlantic Books.

Lin, H.-H., Cao, D.-S., Sethi, S., Zeng, Z., Chin, J. S., Chakraborty, T. S., Shepherd, A. K., et al. (2016). Hormonal modulation of pheromone detection enhances male courtship success. *Neuron* 90: 1272–85.

Lone, S. R., Venkataraman, A., Srivastava, M., Potdar, S., and Sharma, V. K. (2015). *Or47b*-neurons promote male-mating success in *Drosophila*. *Biology Letters* 11. doi: 10.1098/rsbl.2015.0292.

Lowenstein, L. (2002). Fetishes and their associated behavior. *Sexuality and Disability* 20: 135–47.

Lynch, K. S., Rand, A. S., Ryan, M. J., and Wilczynski, W. (2005). Reproductive state influences female plasticity in mate choice. *Animal Behaviour* 69: 689–99.

Madden, J. R., and Tanner, K. (2003). Preferences for coloured bower decorations can be explained in a nonsexual context. *Animal Behaviour* 65: 1077–83.

Magnus, D. (1958). Exerimentelle Untersuchungen zur Bionomie und Ethologie des aisermantels *Argynnis paphia* Girard (Lep. Nymph.). *Zeitschrift für Tierpsychologie,* 15: 397–426.

Malthus, T. (1798). *An Essay on the Principle of Population, as It Affects the Future Improvement of Society with Remarks on the Speculations of Mr. Godwin, M. Condorcet, and Other Writers.* London: Printed for J. Johnson, in St. Paul's Church-Yard.

Marcus, S. (2008). *The Other Victorians: A Study of Sexuality and Pornography in Mid-Nineteenth-Century England.* New Brunswick, NJ: Transaction Publishers.

Marler, P. (1998). Animal communication and human language. In Jablonski, N. G., and Aiello, L. C., eds., *The Origins and Diversification of Language*, 1–19. San Francisco: California Academy of Sciences.

McClintock, M. K. (1971). Menstrual synchrony and suppression. *Nature* 229: 244–45.

McConnell, P. B. (1990). Acoustic structure and receiver response in domestic dogs, *Canis familiaris*. *Animal Behaviour* 39: 897–904.

McCullough, D. (2001). *The Path between the Seas: The Creation of the Panama Canal, 1870–1914.* New York: Simon and Schuster.

Meierjohann, S., and Schartl, M. (2006). From Mendelian to molecular genetics: The *Xiphophorus* melanoma model. *Trends in Genetics* 22: 654–61.

Mello, C., Nottebohm, F., and Clayton, D. (1995). Repeated exposure to one song leads to a rapid and persistent decline in an immediate early gene's response to that song in zebra finch telencephalon. *Journal of Neuroscience* 15: 6919–25.

Menon, V., and Levitin, D. J. (2005). The rewards of music listening: Response and physiological connectivity of the mesolimbic system. *Neuroimage* 28: 175–84.

Meyer, M., Kircher, M., Gansauge, M.-T., Li, H., Racimo, F., Mallick, S., Schraiber, J. G., et al. (2012). A high-coverage genome sequence from an archaic Denisovan individual. *Science* 338: 222–26.

Milinski, M. (2003). Perfumes. In Voland, E., and K. Grammer, K., eds., *Evolutionary Aesthetics*, 325–39. Berlin: Springer.

——— (2006). The major histocompatibility complex, sexual selection, and mate choice. *Annual Review of Ecology, Evolution, and Systematics* 37: 159–86.

Milinski, M., and Wedekind, C. (2001). Evidence for MHC-correlated perfume preferences in humans. *Behavioral Ecology* 12: 140–49.

Miller, G. (2011). *The Mating Mind: How Sexual Choice Shaped the Evolution of Human Nature*. New York: Anchor.

Mitchell, W. B., DiBartolo, P. M., Brown, T. A., and Barlow, D. H. (1998). Effects of positive and negative mood on sexual arousal in sexually functional males. *Archives of Sexual Behavior* 27: 197–207.

Moen, R. A., Pastor, J., and Cohen, Y. (1999). Antler growth and extinction of Irish elk. *Evolutionary Ecology Research* 1: 235–49.

Møller, A. P. (1992). Female swallow preference for symmetrical males. *Nature* 357: 238–40.

Møller, A. P., and Swaddle, J. P. (1997). *Asymmetry, Developmental Stability and Evolution*. Oxford: Oxford University Press.

Møller, A. P., and Thornhill, R. (1998). Bilateral symmetry and sexual selection: A meta-analysis. *American Naturalist* 151: 174–92.

Morton, E. S. (1975). Ecological sources of selection on avian sounds. *American Naturalist* 109: 17–34.

——— (1977). On the occurrence and significance of motivation-structural rules in some bird and mammal sounds. *American Naturalist* 111: 855–69.

Nagel, T. (1974). What is it like to be a bat? *Philosophical Review* 83: 435–50.

Nakano, R., Takanashi, T., Skals, N., Surlykke, A., and Ishikawa, Y. (2010). To females of a noctuid moth, male courtship songs are nothing more than bat echolocation calls. *Biology Letters* 6: 582–84.

O'Connor, J. J., Fraccaro, P. J., Pisanski, K., Tigue, C. C., O'Donnell, T. J., and Feinberg, D. R. (2014). Social dialect and men's voice pitch influence women's mate preferences. *Evolution and Human Behavior* 35: 368–75.

Partridge, L., and Farquhar, M. (1981). Sexual activity reduces lifespan of male fruit flies. *Nature* 294: 580–82.

Pascoal, S., Cezard, T., Eik-Nes, A., Gharbi, K., Majewska, J., Payne, E., Ritchie, M. G., Zuk, M., and Bailey, N. W. (2014). Rapid convergent evolution in wild crickets. *Current Biology* 24: 1369–74.

Pennebaker, J., Dyer, M., Caulkins, R., Litowitz, D., Ackreman, P., Anderson, D., and McGraw, K. (1979). Don't the girls get prettier at closing time? A country and western application to psychology. *Personality and Social Psychology Bulletin* 5: 122–25.

Petrie, M., and Williams, A. (1993). Peahens lay more eggs for peacocks with larger trains. *Proceedings of the Royal Society of London B: Biological Sciences* 251: 127–31.

Pfaff, J. A., Zanette, L., MacDougall-Shackleton, S. A., and MacDougall-Shackleton, E. A. (2007). Song repertoire size varies with HVC volume and is indicative of male quality in song sparrows (*Melospiza melodia*). *Proceedings of the Royal Society of London B: Biological Sciences* 274: 2035–40.

Phelps, S. M., and Ryan, M. J. (1998). Neural networks predict response biases in female túngara frogs. *Proceeding of the Royal Society of London B: Biological Sciences* 265: 279–85.

Prescott, J. W. (2012). Perspective 6: Nurturant versus nonnurturant environments and the failure of the environment of evolutionary adaptedness. In Narvaez, D., Panksepp, J., Schore, A. N., and Gleason, T. R. eds., *Evolution, Early Experience and Human Development: From Research to Practice and Policy*, 427–38. Oxford: Oxford University Press.

Proctor, H. C. (1992). Sensory exploitation and the evolution of male mating behaviour: A cladistic test using water mites (Acari: Parasitengona). *Animal Behaviour* 44: 745–52.

Prosen, E. D., Jaeger, R. G., and Lee, D. R. (2004). Sexual coercion in a territorial salamander: Females punish socially polygynous male partners. *Animal Behaviour* 67: 85–92.

Rodd, F. H., Hughes, K. A., Grether, G. F., and Baril, C.T. (2002). A possible non-sexual origin of mate preference: Are male guppies mimicking fruit? *Proceedings of the Royal Society of London B: Biological Sciences* 269: 475–81.

Rodríguez-Brenes, S., Rodriguez, D., Ibáñez, R., and Ryan, M. J. (2016). Amphibian chytrid fungus spreads across lowland populations of túngara frogs in Panamá. *PLoS One* 11 (5): e0155745.

Rosenthal, G. G., and Evans, C. S. (1998). Female preference for swords in *Xiphophorus helleri* reflects a bias for large apparent size. *Proceedings of the National Academy of Sciences of the United States of America* 85: 4431–36.

Rothenberg, D. (2012). *Survival of the Beautiful: Art, Science, and Evolution*. London: AandC Black.

Ryan, M. J. (1985). *The Túngara Frog: A Study in Sexual Selection and Communication*. Chicago: University of Chicago Press.

——— (1990). Sensory systems, sexual selection, and sensory exploitation. *Oxford Surveys in Evolutionary Biology* 7: 157–95.

——— (2006). Profile: A. Staney Rand (1932–2005). *Iguana* 13: 43–46.

——— (2010). An improbable path. In Drickamer, L., and Dewsbury, D., eds., *Leaders in Animal Behavior: The Second Generation*, 465–96. Cambridge: Cambridge University Press.

——— (2011). Sexual selection: A tutorial from the túngara frog. In Losos, J. B., ed., *In Light of Evolution: Essays from the Laboratory and the Field*, 18–203. Greenwood Village, CO: Ben Roberts and Co.

Ryan, M. J., Bernal, X. E., and Rand, A. S. (2010). Female mate choice and the potential for ornament evolution in túngara frogs, *Physalaemus pustulosus. Current Zoology* 56: 343–57.

Ryan, M. J., Cocroft, R. B., and Wilczynski, W. (1990). The role of environmental selection in intraspecific divergence of mate recognition signals in the cricket frog, *Acris crepitans. Evolution* 44: 1869–72.

Ryan, M. J., and Cummings, M. E. (2013). Perceptual biases and mate choice. *Annual Review of Ecology, Evolution, and Systematics* 44: 437–59.

Ryan, M. J., and Keddy-Hector, A. (1992). Directional patterns of female mate choice and the role of sensory biases. *American Naturalist* 139: S4–S35.

Ryan, M. J., and Rosenthal, G. G. (2001). Variation and selection in swordtails. In Dugatkin, L. A., ed., *Model Systems in Behavioral Ecology*, 133–48. Princeton, NJ: Princeton University Press.

Ryan, M. J., Warkentin, K. M., McClelland, B. E., and Wilczynski, W. (1995). Fluctuating asymmetries and advertisement call variation in the cricket frog, *Acris crepitans. Behavioral Ecology* 6: 124–31.

Salmon, C. (2012). The pop culture of sex: An evolutionary window on the worlds of pornography and romance. *Review of General Psychology* 16: 152.

Schiestl, F. P. (2005). On the success of a swindle: Pollination by deception in orchids. *Naturwissenschaften* 92: 255–64.

Schlupp, I., Marler, C. A., and Ryan, M. J. (1994). Benefit to male sailfin mollies of mating with heterospecific females. *Science* 263: 373–74.

Searcy, W. A. (1992). Song repertoire and mate choice in birds. *American Zoologist* 32: 71–80.

Sedikides, C., Ariely, D., and Olsen, N. (1999). Contextual and procedural determinants of partner selection: Of asymmetric dominance and prominence. *Social Cognition* 17: 118–39.

Seeley, T. D. (2009). *The Wisdom of the Hive: The Social Physiology of Honey Bee Colonies.* Cambridge, MA: Harvard University Press.

Seuss, D. (1988). *Green Eggs and Ham.* New York: Beginner Books / Random House.

Shafir, S., Waite, T. A., and Smith, B. H. (2002). Context-dependent violations of rational choice in honeybees (*Apis mellifera*) and gray jays (*Perisoreus canadensis*). *Behavioral Ecology and Sociobiology* 51: 180–87.

Sigall, H., and Landy, D. (1973). Radiating beauty: Effects of having a physically attractive partner on person perception. *Journal of Personality and Social Psychology* 28: 218.

Silver, N. (2012). *The Signal and the Noise: Why So Many Predictions Fail—but Some Don't.* New York: Penguin.

Simpson, G. G. (1980). *Splendid Isolation: The Curious History of South American Mammals.* New Haven, CT: Yale University Press.

Slotten, R. A. (2004). *The Heretic in Darwin's Court: The Life of Alfred Russel Wallace.* New York: Columbia University Press.

Smith, F. (1990). Charles Darwin's ill health. *Journal of the History of Biology* 23: 443–59.

Steblin, R. K. (2002). *History of Key Characteristics in the Eighteenth and Early Nineteenth Centuries.* Rochester, NY: University of Rochester Press.

Sugiyama, L. S. (2004). Is beauty in the context-sensitive adaptations of the beholder? Shiwiar use of waist-to-hip ratio in assessments of female mate value. *Evolution and Human Behavior* 25: 51–62.

Taylor, C. R., and Rowntree, V. (1973). Temperature regulation and heat balance in running cheetahs: A strategy for sprinters? *American Journal of Physiology—Legacy Content* 224: 848–51.

Taylor, R., and Ryan, M. (2013). Interactions of multisensory components perceptually rescue túngara frog mating signals. *Science* 341: 273–74.

ten Cate, C., and Rowe, C. (2007). Biases in signal evolution: Learning makes a difference. *Trends in Ecology and Evolution* 22: 380–87.

ten Cate, C., Verzijden, M. N., and Etman, E. (2006). Sexual imprinting can induce sexual preferences for exaggerated parental traits. *Current Biology* 16: 1128–32.

Toda, H., Zhao, X., and Dickson, B. J. (2012). The *Drosophila* female aphrodisiac pheromone activates $ppk23^+$ sensory neurons to elicit male courtship behavior. *Cell Reports* 1: 599–607.

Trivers, R. (2011). *Deceit and Self-Deception: Fooling Yourself the Better to Fool Others.* London: Penguin.

Turella, L., Pierno, A. C., Tubaldi, F., and Castiello, U. (2009). Mirror neurons in humans: Consisting or confounding evidence? *Brain and Language* 108: 10–21.

Tuttle, M. (2015). *The Secret Lives of Bats: My Adventures with the World's Most Misunderstood Mammals*. Boston: Houghton Mifflin Harcourt.

Villinger, J., and Waldman, B. (2008). Self-referent MHC type matching in frog tadpoles. *Proceedings of the Royal Society of London B: Biological Sciences* 275: 1225–30.

Vollrath, F., and Milinski, M. (1995). Fragrant genes help Damenwahl. *Trends in Ecology and Evolution* 10: 307–8.

Von Uexküll, J. (2014). *Umwelt und Innenwelt der Tiere*. Berlin: Springer-Verlag.

Waynforth, D. (2007). Mate choice copying in humans. *Human Nature* 18: 264–71.

Weber, E. H. (1978). *E. H. Weber: The Sense of Touch*. Cambridge: Academic Press.

Wedekind, C., Seebeck, T., Bettens, F., and Paepke, A. J. (1995). MHC-dependent mate preferences in humans. *Proceedings of the Royal Society of London B: Biological Sciences* 260: 245–49.

Wilczynski, W., Rand, A. S., and Ryan, M. J. (2001). Evolution of calls and auditory tuning in the *Physalaemus pustulosus* species group. *Brain, Behavior and Evolution* 58: 137–51.

Wiley, R. H. (1973). Territoriality and non-random mating in sage grouse, *Centrocercus urophasianus*. *Animal Behaviour Monographs* 6: 85–169.

Wilkinson, G. S., and Reillo, P. R. (1994). Female choice response to artificial selection on an exaggerated male trait in a stalk-eyed fly. *Proceedings of the Royal Society of London B: Biological Sciences* 255: 1–6.

Winegard, B. M., Winegard, B., and Geary, D. C. (2013). If you've got it, flaunt it: Humans flaunt attractive partners to enhance their status and desirability. *PLoS One* 8: e72000.

Wolf, N. (2003). The porn myth. *New York Magazine*, October 20.

Wyrobek, A. J., Eskenazi, B., Young, S., Arnheim, N., Tiemann-Boege, I., Jabs, E., Glaser, R. L., Pearson, F. S., and Evenson, D. (2006). Advancing age has differential effects on DNA damage, chromatin integrity, gene mutations, and aneuploidies in sperm. *Proceedings of the National Academy of Sciences of the United States of America* 103: 9601–6.

Wyttenbach, R. A., May, M. L., and Hoy, R. R. (1996). Categorical perception of sound frequency by crickets. *Science* 273: 1542–44.

Yeung, C., Anapolski, M., Depenbusch, M., Zitzmann, M., and Cooper, T. (2003). Human sperm volume regulation: Response to physiological changes in osmolality, channel blockers and potential sperm osmolytes. *Human Reproduction* 18: 1029–36.

Yoshizawa, K., Ferreira, R. L., Kamimura, Y., and Lienhard, C. (2014). Female penis, male vagina, and their correlated evolution in a cave insect. *Current Biology* 24: 1006–10.

Zahavi, A. (1975). Mate selection: A selection for a handicap. *Journal of Theoretical Biology* 53: 205–14.

Zahavi, A., and Zahavi, A. (1997). *The Handicap Principle: A Missing Piece of Darwin's Puzzle*. Oxford: Oxford University Press.

Zuk, M., Rotenberry, J. T., and Tinghitella, R. M. (2006). Silent night: Adaptive disappearance of a sexual signal in a parasitized population of field crickets. *Biology Letters* 2: 521–24.

Index